Kids

SUMMER ACADEMY

ARGOPREP

7 DAYS A WEEK

12 WEEKS

- Mathematics
- English
- Science
- Reading
- Writing
- Experiments
- Mazes
- Puzzles
- Fitness

GRADE 8-9

ArgoPrep is one of the leading providers of supplemental educational products and services. We offer affordable and effective test prep solutions to educators, parents and students. Learning should be fun and easy! To access more resources visit us at www.argoprep.com.

Our goal is to make your life easier, so let us know how we can help you by e-mailing us at: info@argoprep.com.

Want much more free worksheets?

Visit us at argoprep.com/worksheets

ISBN: 9781951048754
Published by Argo Brothers.

- ArgoPrep is a recipient of the prestigious **Mom's Choice Award**.
- ArgoPrep also received the 2019 **Seal of Approval** from Homeschool.com for our award-winning workbooks.
- ArgoPrep was awarded the 2019 **National Parenting Products Award, Gold Medal Parent's Choice Award** and **the Tillywig Brain Child Award.**

TABLE OF CONTENTS

TABLE OF CONTENTS

TABLE OF CONTENTS

TABLE OF CONTENTS

HOW TO USE THE BOOK

Welcome to **Kids Summer Academy** by ArgoPrep.

This workbook is designed to prepare students over the summer to get ready for **Grade 9.**
The curriculum has been divided into **twelve weeks** so students can complete this entire workbook over the summer.

Our workbook has been carefully designed and **crafted by licensed teachers** to give students an incredible learning experience.
Students start off the week with English activities followed by Math practice. Throughout the week, students have several fitness activities to complete. Making sure students stay active is just as important as practicing mathematics.
We introduce yoga and other basic fitness activities that any student can complete. Each week includes a science experiment which sparks creativity and allows students to visually understand the concepts. On the last day of each week, students will work on a fun puzzle.

HOW TO WATCH VIDEO EXPLANATIONS

IT IS ABSOLUTELY FREE

Download our app:
ArgoPrep Video Explanations
to access videos on any mobile device or tablet.

or

Step 1 - Visit our website at: www.argoprep.com/k8
Step 2 - Click on the Video Explanations button located on the top right corner.
Step 3 - Choose the workbook you have and enjoy video explanations.

WHAT TO READ OVER THE SUMMER

One of the best ways to increase your reading comprehension level is to read a book for at least 20 minutes a day. We strongly encourage students to read several books throughout the summer. Below you will find a recommended summer reading list that we have compiled for students entering into Grade 9 or simply visit us at: www.argoprep.com/**summerlist**

Author: J. D. Salinger
Title: Catcher in the Rye

Author: William Golding
Title: Lord of the Flies

Author: Matthew Quick
Title: Boy21

Author: Walter Dean Myers
Title: Monster

Author: John Knowles
Title: A Separate Peace

Author: Philip Reeve
Title: Mortal Engines

Author: Alane Ferguson
Title: The Christopher Killer

Author: Neal Shusterman
Title: Scythe

Author: Ernest Cline
Title: Ready Player One

Author: M.T. Anderson
Title: Feed

COMMON CORE TEST SERIES

The goal of these workbooks is to provide mock state tests so students can increase confidence and test scores during actual test day.

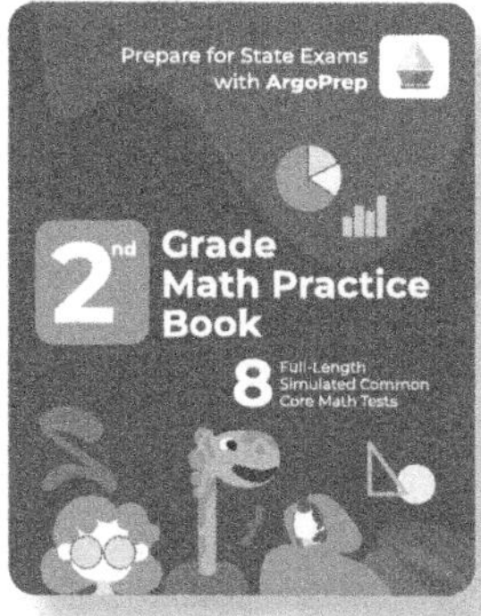

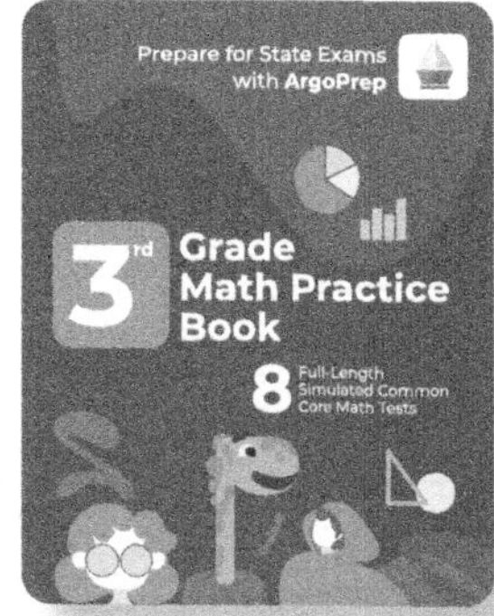

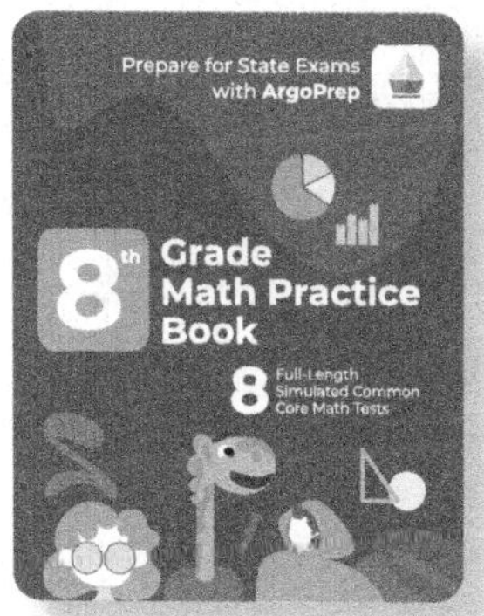

INTRODUCING MATH!

Introducing Math! by ArgoPrep is an award-winning series created by certified teachers to provide students with high-quality practice problems. Our workbooks include topic overviews with instruction, practice questions, answer explanations along with digital access to video explanations. Practice in confidence - with ArgoPrep!

KIDS SUMMER ACADEMY SERIES

ArgoPrep's **Kids Summer Academy** series helps prevent summer learning loss and gets students ready for their new school year by reinforcing core foundations in math, english and science. Our workbooks also introduce new concepts so students can get a head start and be on top of their game for the new school year!

MYSTICAL NINJA

WATER FIRE

ADRASTOS THE SUPER WARRIOR

GREEN POISON

FIRESTORM WARRIOR

RAPID NINJA

CAPTAIN ARGO

THUNDER WARRIOR

CAPTAIN BRAVERY

DANCE HERO

Give your character a name

arieden the destroyer

Write down the special ability or powers your character has and how you will help your community with the powers.

____

____

Great! You are all set. To become an incredible hero, we need to strengthen our skills in **english, math** and **science**. Let's get started.

WEEK 1

GRADE 8-9

Welcome!
Ready to start your amazing journey and strengthen your skills over the summer?
I know I sure am!

Key Terms

Nonfiction - text that is rooted in facts or the truth

Main idea - what the passage or reading is mostly about

Supporting details - sentence details that support and explain the text's main idea

Why is finding the main idea and supporting details in a nonfiction passage so important?

Identifying supporting details in a passage helps explain, support, and prove the main idea. Being able to determine and pinpoint the supporting details and the main idea in a nonfiction passage will allow you to understand and analyze the text.

Let's view an example of how to find supporting details and the main idea in a nonfiction passage!

Read the following passage.

Lately, larger dogs have received negative backlash in the media. Many individuals believe certain breeds of dogs like Pit Bulls, Great Danes, and Rottweilers are dangerous and should not be housed as pets. This preposterous idea is often strengthened by popular opinion, whether truthful or not, when local news channels broadcast stories of dogs biting people or when movies and television shows include dog attacks. Society tends to view these dog breeds as dangerous predators rather than beloved pets. According to this perspective, these dogs are bred for violence and protection, and they are trained to fight other dogs illegally. These negative views give these dog owners and breeders a poor reputation. Sometimes, shelters put these dogs down rather than locate new owners out of fear of the violence they think the dogs possess. However, large dog breeders and pet enthusiasts have finally put these negative views to rest. How pet owners raise a dog determines its violent tendencies, not the breed itself. Researchers have tracked these three different breeds of dogs and recorded incidents over time for their violent tendencies and attacks on community members. The data collected shows that Rottweilers, Great Danes, and Pit Bulls do not violently attack more often than other larger dog breeds like Golden Retrievers and Huskies. The results have surprised people who think negatively about these dogs, but it just goes to show that you cannot believe everything you hear.

What is the main idea of this nonfiction passage? In other words, what is this passage mostly about?

The key to finding the main idea or claim is to focus on what the author is explaining and discussing throughout the **entire** passage. When identifying the main idea, write it out in a complete sentence.

Here is an example:

The author is explaining how certain dog breeds have a bad reputation for violence; however, researchers have proved this theory false.

If the above sentence is the main idea, meaning what the passage is mostly about, then what are the supporting details in the passage?

Supporting details are sentences that explain and support the main idea.

Look at the paragraph again. **What are the sentences that support the main idea?**

One supporting sentence explains how shelters will not take these specific dog breeds. Another supporting detail is how these three dog breeds have a bad reputation. The explanation about how researchers proved this negative reputation wrong would also be a supporting detail in this passage.

Now, it is your turn to try it! Read the short nonfiction passages. Then, find the main idea and supporting details for each.

Rhinos are becoming extinct animals due to poachers. Many individuals in the continent of Africa are poaching rhinos. Countless rhinos are dying each year, which has endangered them and might ultimately lead to their extinction. Why do poachers want to hurt these majestic and beautiful creatures? The reason is their horns provide a high reward. Many poachers receive large sums of money if they kill rhinos and remove their horns to sell. Individuals in many Asian countries purchase the horns and use them in medicines and homeopathic remedies. Many animal activists are trying to save the rhinos in Africa. They are trying to protect the animals by blocking off certain areas and keeping an eye out for poachers.

- What is the main idea of this passage?

....

- What is one supporting detail that supports the main idea?

....

- What is another supporting detail that supports the main idea?

....

In 1930, scientists and astronauts, led by Clyde Tombaugh from the Lowell Observatory in Flagstaff, Arizona, discovered Planet X. They initially called the newly discovered planet, Planet X, because they were unsure of what name to give it. Many of the scientists gave suggestions, but no one could agree on a name. There were many arguments about what to call their miraculous discovery. During the hubbub of debate for a name of the new planet, which has now been demoted to a dwarf planet, one of the scientists mentioned this conundrum at a family gathering. All of a sudden, the scientist's granddaughter suggested the name Pluto. This name struck the scientist's attention. He then went back to work and gave this name suggestion to the other scientists, including Clyde Tombaugh. The decision was made very quickly since Tombaugh fell in love with the name Pluto, which comes from the name of the Roman God of the underworld.

- What is the main idea of this passage?

- What is one supporting detail that supports the main idea?

- What is another supporting detail that supports the main idea?

At night, when gazing high up at an electrical tower, one might notice a blinking light. Do you know the reason why there are lights up on tall towers? These bright lights warn planes so they do not hit them when flying low or when getting ready to land. Makes sense, right? However, did you know that those lights were intially only solid beams? Later, regulations were put in place to make the lights blink. Is the reason for the change so planes can see them better? Not exactly! The reason why the lights blink rather than stay on steadily is because birds were actually crashing into them. According to the Federal Aviation Administration, the solid lights were attracting the birds and disorienting them, and many lost their lives after crashing into the electrical towers. By mandating the change to blinking lights, planes and birds are now safe.

- What is the main idea of this passage?

- What is one supporting detail that supports the main idea?

- What is another supporting detail that supports the main idea?

Let's get some fitness in! Go to page 237 to try some fitness activities.

Part I - Allusion

Key Terms

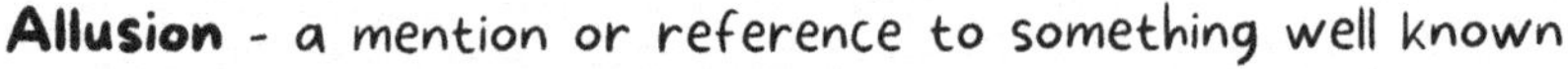

Allusion - a mention or reference to something well known

What are allusions?

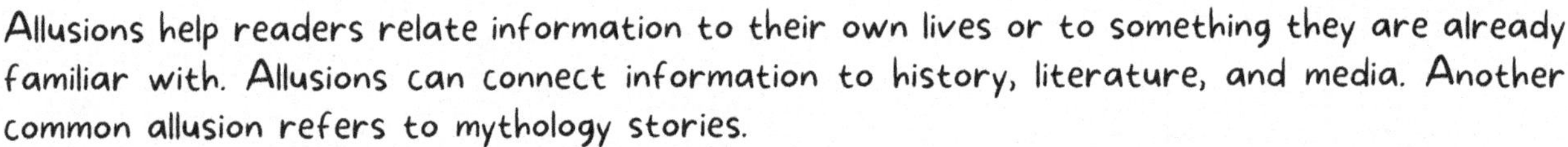

Allusions help readers relate information to their own lives or to something they are already familiar with. Allusions can connect information to history, literature, and media. Another common allusion refers to mythology stories.

Here is an example that contains an allusion:

In a television show, the main character is talking to her best friend and tells her that she "better get home because it is late and she may turn into a pumpkin!" While this line may be humorous, it is also an allusion.

What is this example alluding to? What is its meaning?

This example alludes to the famous fairy tale, *Cinderella*. The main character, Cinderella, had to hurry home before midnight because that is when the magic from her fairy godmother would run out. Cinderella had to hurry home so people at the ball, especially Prince Charming, did not know the truth about the magic and discover she was really poor. This allusion in the television show tells the audience that the character has to hurry home before midnight, just like Cinderella did.

Let's practice!

Below there are going to be some sentences that contain allusions. Identify the allusion and explain what the allusion means.

1. This year's football championship was definitely a David and Goliath story. To everyone's surprise, the Elkhart Titans defeated the West Side Jaguars.

What is the allusion in this sentence? ....

What does this allusion mean? ....

2. We thought this was not going to end well, but to our surprise, the Good Samaritan ran into the road and grabbed our little puppy, Eli, before the semi truck could hit him.

What is the allusion in this sentence? ....

What does this allusion mean? ....

3. Mrs. Rosen allowed her daughter, Tia, to pick the name for their new kitten they adopted at the Humane Society. Tia thought about it for a while and finally decided to name her new kitten, Kitty. Mrs. Rosen thought this name was very odd and common, but she decided that Kitty was a rose by any other name, and it was just as sweet.

What is the allusion in this sentence? ....

What does this allusion mean? ....

Read the mythology story below. Then, answer the questions about allusions that follow.

Story of King Midas

King Midas had everything a king could ever wish for. He lived in a luxurious castle and shared his life of abundance with his beautiful daughter. Even though King Midas was extremely rich, richer than anyone else in the world, he wanted more. King Midas spent his time counting his gold, and soon, his avarice got the best of him. He would often take baths in his gold coins and sleep with them at night. One day, King Midas came across a god who granted him a wish. King Midas did not have to think too hard. He immediately wished that everything he touched turned to gold.

While King Midas thought this wish would be what he had always wanted, it turned out to be a curse. When King Midas came home, he stepped inside his mansion and the floor turned to gold. He sat down to dinner, and his dinner table and his food turned to gold. When his daughter came into the house after school, he hugged her, and his daughter turned into a golden statue.

4. What is the meaning of the following allusion? - "When it comes to growing crops in her garden, Lucile had the Midas touch."

....

5. Write your own allusion, referencing the mythology story of King Midas. ....

....

Part 2 - Point of View

Key Terms

First-Person Point of View - the story is told by a particular character in the story

- Keywords in first-person point of view: I, me, my, we, us, and our

Second-Person Point of View - the story allows the reader to be a part of it

- Keywords in second-person point of view: you and your

Third-Person Point of View (Limited) - the story is being told by an outside narrator who is not a character in the story

- Keywords in third-person point of view (limited): he, she, her, him, and they

Third-Person Point of View (Omniscient) - the story is being described by an outside narrator; however, the narrator is all-knowing, almost godlike when describing the characters' thoughts and feelings

- Keywords in third-person point of view (omniscient): he, she, her, him, and they

Read the following passages and identify the point of view being used. Then explain why it is that specific point of view.

1. "We've got Father and Mother and one another," said Beth. Beth was the shy one. She always looked on the bright side of things. "We haven't got Father," said Jo sadly. "And we might not have him for a very long time." The girls were suddenly silent. (From *Little Women* by Louisa May Alcott - Public Domain Material)

....

....

2. My grandpa Ellis was my favorite person in the entire universe. He was my mother's father, stepfather to be exact, but he treated her like his own. He was always proud of me for everything I did. I will miss him dearly.

....

....

3. It is April, 1967 and you are fourteen years old, walking into high school for the first time. You look around, and no one looks like you. You take a big gulp and push your fears down to your empty stomach. You hope this school year goes well.

....

....

4. The bus picks up the intermediate students every afternoon at a quarter past three. A little girl with red hair always stands by herself at the bus stop and always sits alone on the bus.

....

....

Let's get some fitness in! Go to page 237 to try some fitness activities.

Rational and Irrational Numbers

Review:

A rational number is an integer, fraction, or decimal. It can be positive or negative. In its decimal form, a rational number must be finite or have a repeating pattern behind the decimal.

Question 1:

Which number is irrational?

A. $\frac{1}{3}$

B. 333

C. -3

D. $\sqrt{17}$

Question 2:

Which number is irrational?

A. $1\frac{5}{9}$

B. - 0.375

C. $\frac{6}{7}$

D. π

Question 3:

Which number is irrational?

A. $\sqrt{15}$

B. $\frac{1}{7}$

C. $\frac{2}{7}$

D. $\frac{3}{7}$

Question 4:

Which number is rational?

A. 2.5

B. $2\frac{1}{2}$

C. -2.5

D. All of the above

Question 5:

Captain Argos is impressed by his flying speed over the ocean. In miles per hour, he's calculated his speed to be the same as π. Is this speed a rational or irrational number? Explain.

Question 6:

On the beach, Captain Argos proudly lifts a dozen beach balls weighing a total of $3\frac{1}{3}$ kilograms. Is this weight a rational or irrational number? Explain.

Question 7:

To show off his abilities, Captain Argos leaps over a sandcastle 0.62 meters high. Is this height a rational or irrational number? Explain.

Question 8:

Captain Argos takes a break after spending the morning saving starfish stranded on the beach. He sunbathes for $\frac{19}{2}$ hours. Is this a rational or irrational number? Explain.

Let's get some fitness in! Go to page 237 to try some fitness activities.

Defining Numbers in Mathematical Terms

Review:

A whole number is a positive number that is not expressed as a fraction or decimal. An integer is a positive or negative number which is not expressed as either a fraction or decimal. A fraction is expressed as the ratio of two integers. A decimal is expressed as an integer with the fractional part behind the decimal.

Question 1:

How would you define 365? Circle all that apply.

A. A whole number
B. An integer
C. A fraction
D. A decimal
E. A rational number
F. An irrational number

Question 2:

How would you define $1\frac{7}{8}$? Circle all that apply.

A. A whole number
B. An integer
C. A fraction
D. A decimal
E. A rational number
F. An irrational number

Question 3:

How would you define the "golden ratio" in mathematics (1.6180339887...)?

A. A whole number
B. An integer
C. A fraction
D. A decimal
E. A rational number
F. An irrational number

Question 4:

How would you define π in mathematics (3.141592653...)?

A. A whole number
B. An integer
C. A fraction
D. A decimal
E. A rational number
F. An irrational number

Question 5:

Captain Argos screamed when he saw a group of 52 lobsters moving towards him. Using the mathematical terms described in the review, how would you define this number? Is it rational or irrational?

Question 6:

During a snorkeling adventure, Captain Argos spotted an eel with a length of $\frac{1}{2}$ meters long and quickly swam in the other direction. How would you define this number? Is it rational or irrational?

Question 7:

While scuba diving, Captain Argos got tangled in his cape and sunk to a depth of -10 meters. How would you define this number? Is it rational or irrational?

Question 8:

Captain Argos tells the tale of saving a baby dolphin with a total weight in pounds equal to 5 times π. How would you define this number? Is it rational or irrational?

Let's get some fitness in! Go to page 237 to try some fitness activities.

Ordering Numbers in Various Forms

Question 1:

Order these numbers from least to greatest: $\frac{1}{2}, \frac{1}{4}, \frac{2}{3}, \frac{5}{8}$

..........., ..., ..., ...

Question 3:

Order these numbers from least to greatest: $\frac{1}{5}, \frac{1}{6}, \frac{1}{7}, \frac{1}{8}$

..........., ..., ..., ...

Question 2:

Order these numbers from least to greatest: 0.6, -0.6, 0.65, -0.65

..........., ..., ..., ...

Question 4:

Order these numbers from least to greatest: -3, π, 0.3, 3

..........., ..., ..., ...

Question 5:

Captain Argos thinks he's seen a dangerous looking flounder. To hide, he has sunk to various depths in the water. He has made note of them: -15 feet, -25 feet, -5 feet, -10 feet. Place these distances on the number line:

Question 6:

Captain Argos has rescued several tiny clams out of a tidepool. He notes the diameters of their shells: $5\frac{1}{4}$ centimeters, $3\frac{3}{4}$ centimeters, 5.75 centimeters, and 3.25 centimeters. Place these measurements on the number line:

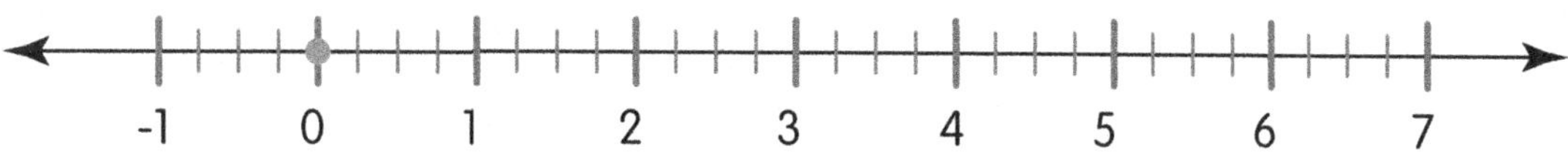

Question 7:

Captain Argos has survived an allergic reaction to shrimp and has lived to tell the tale. He describes the random numbers that drifted through his mind during his delirium: 0.8, 8, $\frac{1}{8}$, and -8. Place these numbers on the number line:

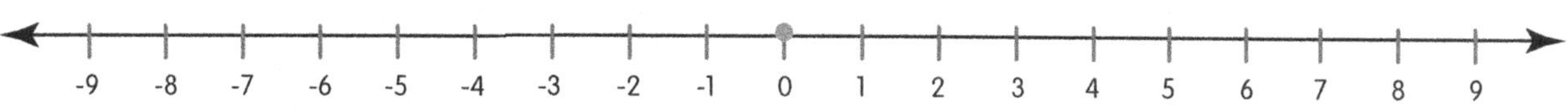

Question 8:

Following his death-defying encounter with shrimp, Captain Argos decides he is done eating shellfish. He heads to the hot dog stand on the beach for a 12-inch hot dog in a $13\frac{1}{2}$ inch bun. Onto it, he squirts 6.8 inches of ketchup and 8.6 inches of mustard. Place all four of these numbers on the number line:

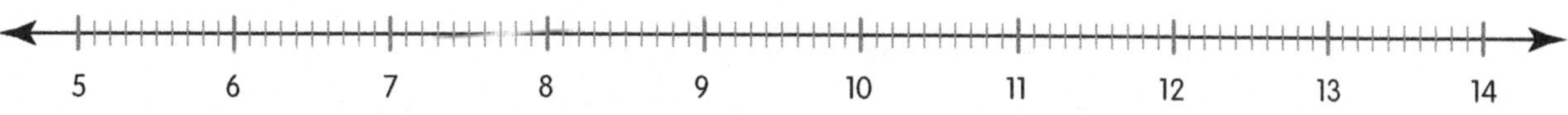

Let's get some fitness in! Go to page 237 to try some fitness activities.

What is Diffusion?

Diffusion is the exchange or the intermingling of substances. It is the shifting of particles through the process of natural movement. Diffusion happens in everyday life. For example, if you ever spray perfume or cologne, you are diffusing something. The liquid spray diffuses into the air and goes into your nose. Another example of daily diffusion includes a tea bag placed in hot water. The tea leaves diffuse in the scalding water to make a drinkable substance. The word diffusion comes from Latin and means "to spread out." Like in the above examples, the perfume or cologne spreads out when sprayed, and the tea leaves spread out into the hot water.

In today's experimental protocol, meaning the steps involved in an experiment, you will observe the process of diffusion through food coloring and water in order to see how the color spreads. This specific experiment will explain how temperature affects the process of diffusion.

Materials Needed

2 glass beakers	200 milliliters of hot water	200 milliliters of cold water
Food coloring	Stopwatch	

Make A Prediction

1. Do you think the temperature of the liquid matters in the diffusion of the food coloring? Will the temperature of the water make the diffusion faster or slower? Explain.

....

Procedure

1. Pour 200 milliliters of hot water into one of the beakers.
2. Place three drops of food coloring into the beaker of hot water.
3. Using the stopwatch, measure how long it takes the food coloring to diffuse through the water, turning the water another color.
4. Record the time : ....
5. Pour 200 milliliters of cold water into another beaker.
6. Place three drops of food coloring into the beaker of cold water.
7. Using the stopwatch, measure how long it takes the food coloring to diffuse through the water and turn the water another color.
8. Record the time : ....

Diffusion

SCIENCE EXPERIMENT

Questions

2. Were you correct in your prediction? Explain.

....

....

3. Which temperature of water did the food coloring diffuse in the fastest?

....

....

4. Based on your results, how does this explain the correlation between temperature and diffusion?

....

....

5. What if the food coloring was dropped into 200 milliliters of cold water and in 400 milliliters of hot water. Explain how you think the results would be different from the first experiment.

....

....

6. In a paragraph, explain how the temperature of water affects diffusion and analyze the results of the experiment.

....

....

Let's get some fitness in! Go to page 237 to try some fitness activities.

MAZE

Directions: The wires are a mess! Help music lovers by connecting the numbered headphones with the lettered music players.

1 ____ 2 ____ 3 ____ 4 ____

1

2

A

B

C

4

3

D

WEEK 2

GRADE 8-9

Nice Work!
You are ready to move to Week 2. We will learn about the author's purpose, tone, and figurative language. We will also practice exponents in detail.

1 2 3 4

Part I - Author's Purpose

Key Terms:

Author's Purpose - what the author hopes to achieve by writing

What are the different reasons why authors choose to write?

An author's purpose is why an author chooses to write to a particular audience and why they are writing the passage in the first place. The author's purpose can be determined by looking at the diction (word choice) the author uses, as well as the tone or the author's attitude. There are three main purposes for writing.

1. **To Persuade**

Some authors want to persuade their readers. This style is argumentative and makes the reader think, feel, or react to what the author is writing about. They want their readers to hold the same position as they do.

2. **To Inform or to Explain**

Many authors want their readers to learn something, so they try to inform them or teach them about a particular topic. To inform an audience, authors use facts, statistics, and research.

3. **To Describe or to Entertain**

Rather than persuading you to believe something or teach you something, a writer's sole purpose may just be to entertain their audience. To do this, they will usually use descriptive diction, figurative language, and sensory details so the audience can visualize what they are reading.

The reason why authors write depends on who their audience is. They know who they are writing to, so they structure their writing for a particular group of people in order to achieve a desired outcome.

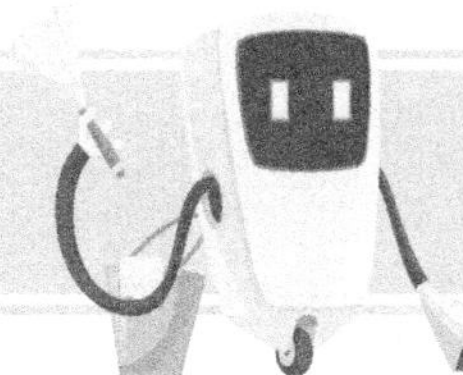

Read the following passages. Then determine the audience and the author's purpose in each passage.

1. The SouthSide Nuggets basketball team endured a tough opening game to their season with a 23 to 89 loss against the NorthSide Penguins. The Penguins are seated third in the state and first in the conference. The Nuggets have not lost since two seasons ago when they played the Penguins previously. Coach Rheinhold noted, "This was a tough game, and the team has some learning to do in order to get better."

....

....

....

2. West High School will put on the classic play *Romeo and Juliet* next Friday in the Reimer Auditorium. This Shakespeare drama will make audiences of all ages laugh and cry. There are over one hundred students participating in the winter play this year, and they would love the community's support.

....

....

....

3. To whom it may concern:
Why is the library decreasing their hours when they should be increasing their hours? I know the library feels that this will cut costs by shutting down earlier; however, doing so will hurt the students in this community. They go to the library to find books, to study, and to use the internet for homework, assignments, and projects. For some, the library is the only outlet for a safe and quiet place to study and a connection to strong, fast internet. Reducing library hours will hurt students who just want to be stronger academically.

....

....

....

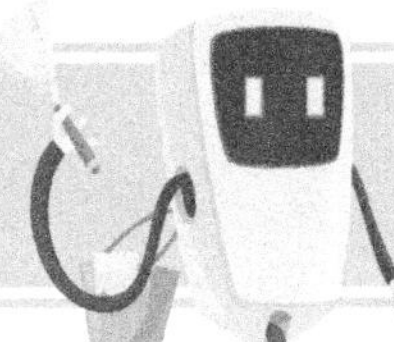

Part 2: Tone

Key Term:

Tone - the author's attitude or feeling that is expressed in a text

How do I identify the tone in a reading?

We can identify the tone by looking at the words or the diction the author chooses. These words will reveal the feelings behind them and the author's attitude when writing the piece.

Here is an example of how to identify tone:

Look at the following sentence:

As I walked home from school and passed the graveyard, the dark trees were swaying violently.

Notice the words "graveyard," "dark," and "violently." All three of these words give off a foreboding connotation and attitude. This sentence has a dark and eerie tone.

Read the following poems. Then determine the tone of the poem and support your answer.

To a Child Dancing Upon the Shore

by William Butler Yeats

Dance there upon the shore;
What need have you to care
For wind or water's roar?
And tumble out your hair
That the salt drops have wet;
Being young you have not known
The fool's triumph, nor yet
Love lost as soon as won.
And he, the best warrior, dead
And all the sheaves to bind!
What need that you should dread
The monstrous crying of wind?

....

....

A Dream

by Stephen Phillips

My dead love came to me, and said: 'God gives me one hour's rest,
To spend with thee on Earth again: How shall we spend it best?'
'Why, as of old,' I said; and so
We quarrelled, as of old: But, when I turned to make my peace,
That one short hour was told.

....

....

Let's get some fitness in! Go to page 237 to try some fitness activities.

What is Figurative Language?

Remember that figurative language is language or wording that is not meant to be taken literally. Authors use figurative writing in order to express themselves.

Key Terms:

Verbal Irony - saying one thing but meaning something completely different

- Example - Tyra was thrilled she had another math test today, since she forgot to study.
- Explanation - This is an example of irony because even though Tyra said she was "thrilled" that she had a math test, she did not really mean that. She was actually the opposite of thrilled.

Simile - comparing unlike things using the words "like" or "as"

- Example - Her hands were as cold as ice.
- Explanation - This is an example of a simile because the author is comparing the girl's hands to ice. The sentence also contains the word "as" to compare.

Metaphor - comparing unlike things

- Example - Her smile is happiness.
- Explanation - This is an example of a metaphor because the author compared the girl's smile to the state of happiness. Note that the sentence did not use the words "like" or "as." That is why it is a metaphor and not a simile.

Idiom - a nonliteral expression; usually well known

- Example - It's raining cats and dogs.
- Explanation - This is an idiom. This sentence is an example of a well known saying that should not be taken literally. People tend to say "It's raining cats and dogs" to indicate or explain it is raining really hard.

Personification - giving non-human objects human qualities

- Example - The trees were waving in the breeze.
- Explanation - This is personification because the trees are being described as completing a human action like "waving." Trees cannot wave like people do because they do not have arms and hands.

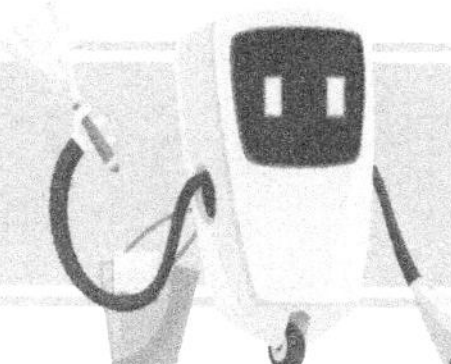

Onomatopoeia - sound words or any words that describe or characterize a sound

Example - The soup was popping in the bowl.

Explanation - The word "popping" is an example of onomatopoeia because it is a sound word.

Alliteration - repetition of initial consonant sounds (consonant sounds are any letters that are not vowels)

Example - Todd the turkey took a taco.

Explanation - This sentence is an example of alliteration because the letter "t" repeats in the words "Todd," "turkey," "took," and "taco."

Read the sentences below. Then, identify the figurative language and explain the meaning of it when used in the sentence.

NOTE - some sentences may have multiple examples of figurative language.

1. The family reunion was enjoyable. I felt like I was getting three teeth pulled at the dentist.

What is the figurative language used? ....

What does it mean? ....

2. Jill had a great night's sleep. Her baby snored as quiet as a car alarm.

What is the figurative language used? ....

What does it mean? ....

3. "Today is the best day ever," shouted Valarie after her car broke down, her coffee spilled, and her boss fired her.

What is the figurative language used? ....

What does it mean? ....

4. Her laughter is hope. Her eyes are in despair.

What is the figurative language used? ....

What does it mean? ....

5. It is up to Paul. He needs to make the decision. The ball is in his court.

What is the figurative language used? ....

What does it mean? ....

....

6. After Paula poured the leftover milk down the drain, the sink sighed.

What is the figurative language used? ....

What does it mean? ....

....

7. The military separates the men from the boys.

What is the figurative language used? ....

What does it mean? ....

....

8. The stew in the scalding pot was bubbling, snapping, and cracking very loudly.

What is the figurative language used? ....

What does it mean? ....

....

9. Mrs. Benson bought a very expensive purse she saw in Baton Rouge, Louisiana during her last vacation.

What is the figurative language used? ....

What does it mean? ....

....

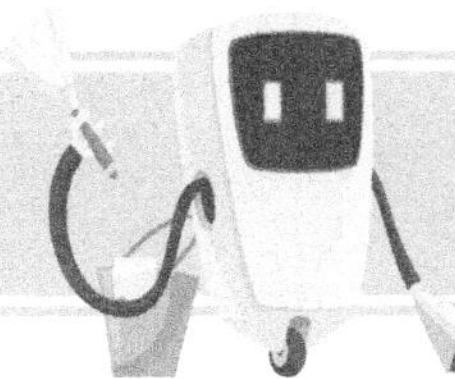

Write complete sentences that contain figurative language.

10. Give an original example of verbal irony.

....

....

11. Give an original example of a simile.

....

....

12. Give an original example of a metaphor.

....

....

13. Give an original example of an idiom.

....

....

14. Give an original example of personification.

....

....

15. Give an original example of onomatopoeia.

....

....

16. Give an original example of alliteration.

....

Let's get some fitness in! Go to page 237 to try some fitness activities.

Simplifying the Product of Integers with Exponents

Review:

When multiplying the same base integer with positive or negative exponents, simplify by finding the sum of the exponents. Keep the base integer the same.

Question 1:

Simplify $10^{8} \times 10^{12}$

A. 10^{20}
B. 10^{4}
C. 100^{20}
D. 10^{-4}

Question 2:

Simplify $5^{10} \times 5^{6} \times 5^{-2}$

A. 125^{14}
B. 15^{-14}
C. 5^{14}
D. 5^{18}

Question 3:

Simplify $3^{-8} \times 3^{-4} \times 3^{-2}$

A. 27^{-14}
B. 9^{-14}
C. 3^{-64}
D. 3^{-14}

Question 4:

Simplify $7^{-1} \times 7^{1} \times 7^{1}$

A. 21^{3}
B. 7^{1}
C. 7^{-1}
D. 7^{3}

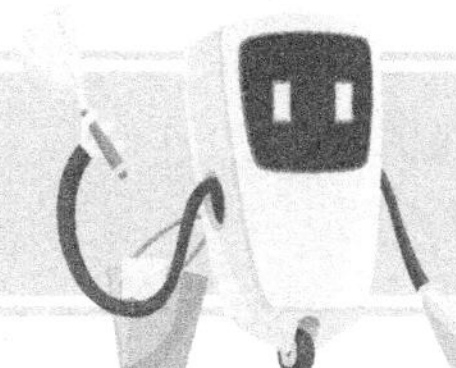

Question 5:

Mystical Ninja meditates on the beach by counting the waves in exponential form. His wave count is $8^2 \times 8^5 \times 8^3$. What is the simplified form?

Question 6:

Focusing on the grains of sands where he's sitting, Mystical Ninja makes an immediate count of them in exponential form. His count is $12^{100} \times 12^{100} \times 12^{100} \times 12^{100}$. What is the simplified form?

Question 7:

Mystical Ninja sees the size of a micro-organism floating in the water in exponential form. His result in millimeters is $9^{-9} \times 9^{-9}$. What is the simplified form?

Question 8:

Mystical Ninja feels wind currents build up and die away. He calculates the increase and decrease in their speed in exponential form. His count is $3^2 \times 3^{-2} \times 3^5$. What is the simplified form?

Let's get some fitness in! Go to page 237 to try some fitness activities.

Simplifying the Quotient of Integers with Exponents

Review:

When dividing the same base integer with positive or negative exponents, simplify by subtracting the exponent in the denominator from the exponent in the numerator. Keep the base integer the same. (Remember: in a division problem, the numerator appears first and the denominator appears second.)

Question 1:

Simplify $5^{12} \div 5^{10}$

A. 5^{2}

B. 1^{2}

C. 5^{22}

D. 5^{-2}

Question 2:

Simplify $4^{5} \div 4^{2}$

A. 4^{7}

B. 1^{3}

C. 4^{3}

D. 4^{-3}

Question 3:

7^{3} simplifies which expression?

A. $7^{7} \div 7^{10}$

B. $7^{10} \div 7^{7}$

C. $7^{3} \div 7^{3}$

D. None of the above

Question 4:

10^{15} simplifies which expression?

A. $10^{5} \div 10^{10}$

B. $10^{10} \div 10^{5}$

C. $10^{3} \div 10^{5}$

D. None of the above

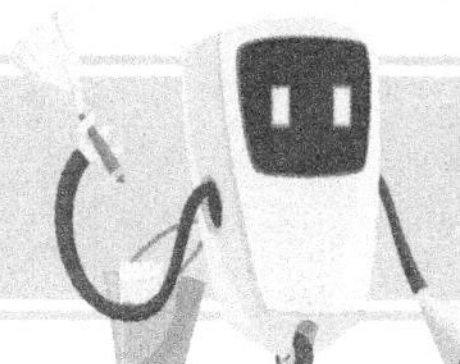

Question 5:

Mystical Ninja can clearly see the simplification of integers with exponents. For $9^8 \div 9^7$, what does he see as the simplification?

Question 6:

Mystical Ninja knows the simplification of $6^{20} \div 6^{15}$. What is it? How do you know?

Question 7:

Mystical Ninja believes the simplification of $8^{100} \div 8^{50}$ is 8^{50}. Is he correct? Explain.

Question 8:

Mystical Ninja believes the simplification of $100^{100} \div 100^{98}$ to be 100. Is he correct?

Let's get some fitness in! Go to page 237 to try some fitness activities.

Negative Integers with Exponents

Review:

With exponents, when the base integer is a negative number, then you are multiplying a negative number. When the base integer is a positive number with a negative sign lying outside of parenthesis, the final answer is considered negative.

Question 1:

Solve -6^2

A. -36
B. -12
C. 12
D. 36

Question 2:

Solve $-(12^2)$

A. -144
B. 144
C. 24
D. -24

Question 3:

Which expression is equivalent to -25?

A. $-(5^2)$
B. -5^2
C. 5^2
D. 5^{-2}

Question 4:

Which expression is equivalent to 100?

A. -10^2
B. 10^2
C. $-(10^2)$
D. Both A and B

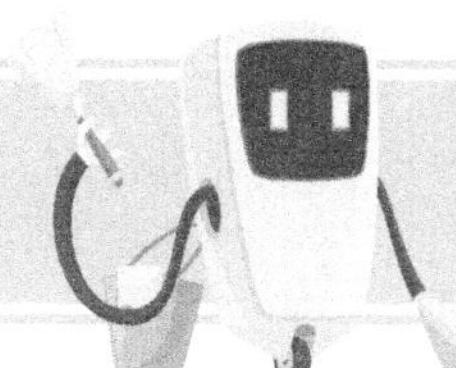

Question 5:

Hoping to avoid negative energy, Mystical Ninja solves the expression -9^4. What is the answer?

Question 6:

Feeling the power of the ocean's undertow, Mystical Ninja solves the expression -5^3. What is the answer?

Question 7:

Mystical Ninja relies on positive thinking when he observes -81 scratched in the sand. What integer with an exponent would lead to this answer?

Question 8:

One evening, Mystical Ninja lies back on the sand and contemplates the -8 written in the stars. What integer with an exponent would lead to this answer?

Let's get some fitness in! Go to page 237 to try some fitness activities.

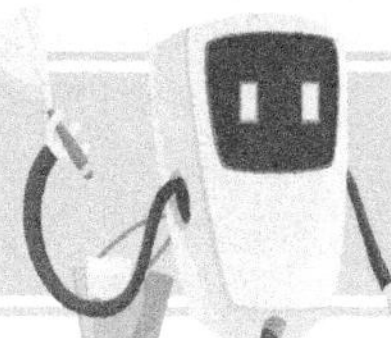

The Greenhouse Effect

Have you ever entered a greenhouse? If you have, you know that when you are inside, it is very hot. The glass walls of the greenhouse keep and trap the heat in because this is the best environment to grow plants and flowers.

Many people are concerned by the harm the greenhouse effect has on our environment.The effect is caused by gases entering into our atmosphere. The problem is that these gases remain, and the atmopshere traps the Earth inside just like the glass traps heat. The gasses that are stored in our atmosphere are carbon dioxide (CO_2), nitrous oxide (N_2O), methane (CH_4), and various fluorinated gases. While the sun shines, light is the only element that can pass through. The gasses have no way to escape. These trapped gasses then begin to heat up our atmosphere.

If the greenhouse gasses were not here, the temperature of Earth would drop, but due to their presence, they are heating up the Earth and causing major problems. While there are some natural causes for the greenhouse effect, there are many intentional and avoidable ones as well. Some natural causes of Carbon Dioxide in the atmosphere are the byproducts of respiration from humans and animals. Methane exists in our atmosphere naturally due to the decomposition of plants and animals, the presence of landfills, and the release of gasses from volcanoes. While these are reasons why some of these gasses occur naturally in our atmosphere, they are not the only reasons for their presence. The excess of trapped gasses is increasing rapidly.

Since the early 1970s, greenhouse emissions have increased by seventy percent. The major reason is the elevation of carbon dioxide, or CO_2. This excess gas in our atmosphere comes from using and burning fossil fuels. Fossil fuels are also released by automobiles, as well as by trains, planes, and power plants. The extra presence of methane and nitrous oxide in our atmosphere is due to the burning of fossil fuels such as coal. Fluorinated gases such as Chlorofluorocarbons (CFCs) are the result of the incorrect disposal of refrigerators and aerosol cans like hairspray.

When these added gases are trapped in the atmosphere, the temperature of Earth increases, adding to global warming. The increased temperatures are causing the glaciers to melt at a much faster rate. This is a problem because it adds more water to the oceans, elevating sea levels and leading to flooding. Some people living in and working in low level areas have to move. Island populations, as well as people living in coastal areas, such as Florida, suffer also. Trapped gasses also contribute to changes in precipitation. Some areas are seeing increased rates, while others are receiving no precipitation at all, both of which drastically affect animals and crops.

People want to decrease the amount of greenhouse gases, but how do we do it? Some main ways include using public transportation, recycling, planting more trees, and supporting other means of energy rather than using fossil fuels.

1. Why is this phenomenon known as the greenhouse effect?

2. Are the natural causes of greenhouse gases a problem? Explain.

3. Why have greenhouse emissions increased so much since the 1970s?

4. How is the increase and decrease of precipitation in certain areas a problem? How does this affect crops?

5. What other means of energy should we support instead of fossil fuels? Explain.

Let's get some fitness in! Go to page 237 to try some fitness activities.

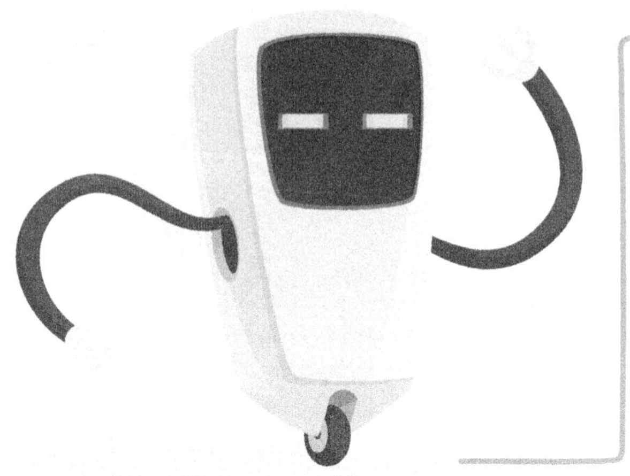

MAZE GAME

ARGOPREP
WEEK 3
GRADE 8-9
Practice makes progress!
In Week 3 we will learn about how to identify themes in reading passages. For math, we will work on squares, cubes, and square roots. There's also a fun science experiment to learn about the freezing point of water!

What is a theme and how do you find it?

In fiction writing, a theme is the key message the author wants the audience to learn after reading a text. In one text there can be multiple themes. Be careful, though; some readers think the theme is what the text is about. That is the main idea. The theme is what the reader learns after reading the text.

The theme is not told to you outright. You need to read the entire text to then think about what the author wants you to learn. Then, you need to write the theme in a sentence.

Here is an example of how to find the theme in a reading:

Look at the quote from Natalie Babbitt's novel *Tuck Everlasting* (Public Domain Material): "She was afraid to go alone. It was one thing to talk about being by yourself, doing important things, but quite another when the opportunity arose."

In order to find the theme, you need to analyze the text and dig deep into the diction. What does the author want you to learn from these lines? What value can be found in these words? What lesson could someone take with them and use in their life after reading these lines? What about the lesson of bravery? It is one thing to talk about being brave, it is completely different when you are forced into a situation that requires bravery.

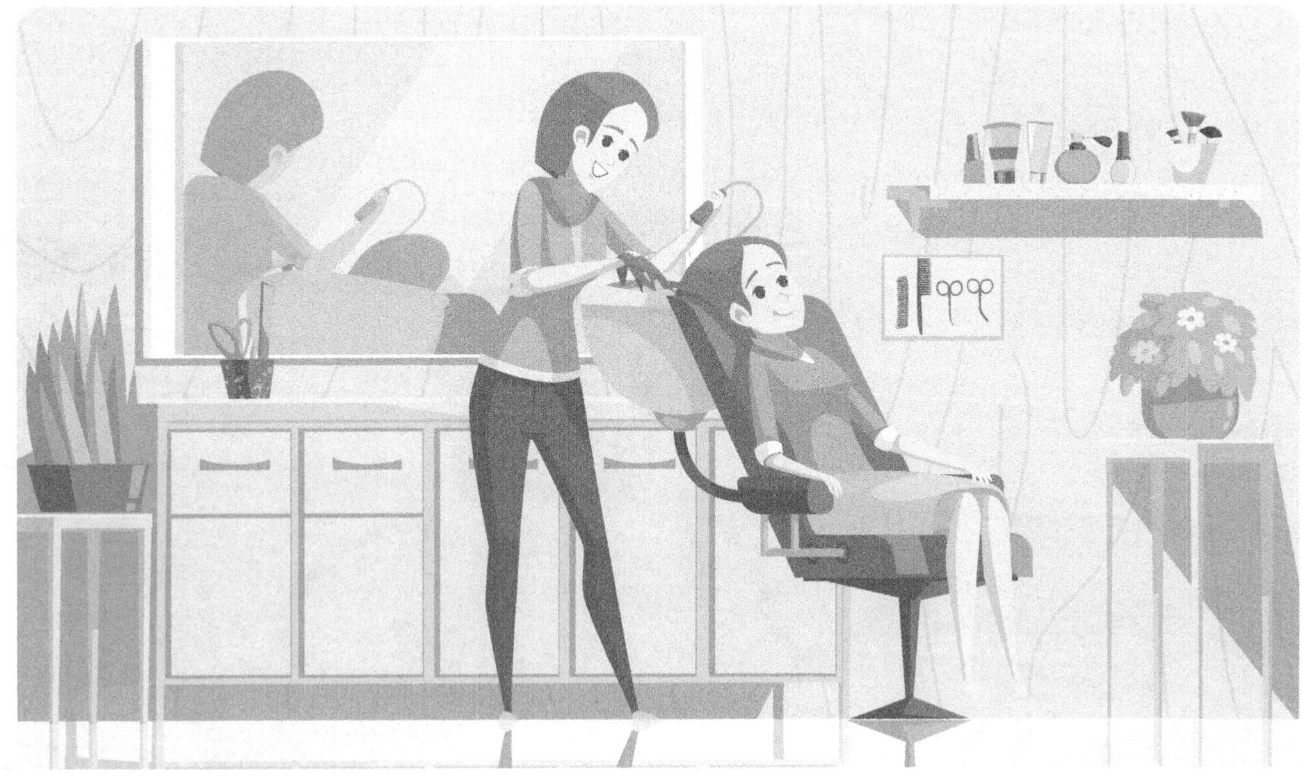

Let's practice! Read the short story below. Then identify the theme of the story.

The Gift of the Magi

By O. Henry

One dollar and eighty-seven cents. There was clearly nothing to do but flop down on the shabby little couch and howl. Tomorrow would be Christmas Day, and Della had only $1.87 with which to buy Jim, her husband, a present. She pulled down her hair and let it fall to its full length below her knees. All of a sudden, a smile appeared on her face. She grabbed her coat and went out the door. She walked until she stopped at a sign that read: "Hair Goods of All Kinds." She ran in and asked, "Will you buy my hair?" The lady looked at her hair. "Twenty dollars," the lady answered.

After Della received her money, she went into a store and knew exactly what she wanted to buy for Jim. It was a platinum fob chain simple for his watch--the watch that he loved so much. She didn't like her new short hair and thought Jim would think she looked like a boy, but she was happy she could give him such a nice gift.

At 7 o'clock, she heard her husband walk up to the door. The door opened and Jim stepped in. His eyes were fixed upon Della, and there was an expression in them she could not read, and it terrified her.

"Jim, darling," she cried, "don't look at me that way. I had my hair cut off and sold because I couldn't have lived through Christmas without giving you a present."

"You've cut off your hair?" asked Jim.

"I'm me without my hair, aren't I?" said Della.

Jim drew a package from his overcoat pocket and threw it upon the table. "There's not anything in the way of a haircut that could make me like my girl any less. But if you'll unwrap that package, you may see why I was surprised."

Della opened the package, and there lay a set of hair combs, beautiful ones made of pure tortoise shell with jewelled rims. She had hair combs but no hair. She wanted to cry.

Then Della gave Jim her gift. "Isn't it dandy, Jim? I hunted all over town to find it. Give me your watch. I want to see how it looks with the chain."

"Della," said Jim. "I don't have my watch anymore. I sold the watch to get the money to buy your combs."

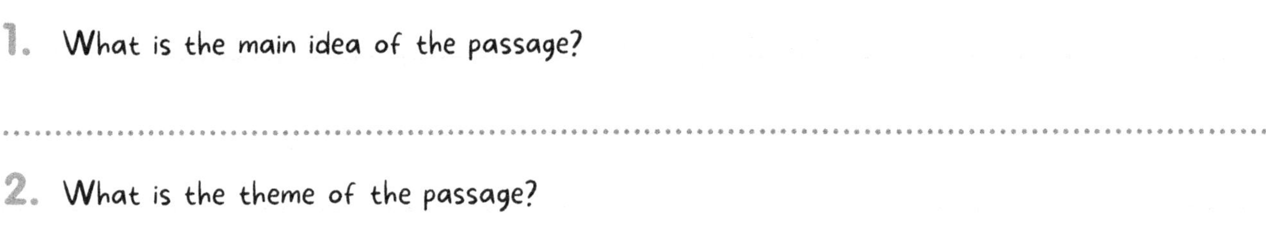

1. What is the main idea of the passage?

....

2. What is the theme of the passage?

....

3. Identify an instance of figurative language used in the passage and describe its meaning.

....

4. Identify another instance of figurative language used in the passage and describe its meaning.

....

Read the poem below. Then, identify the theme.

Remember

by Christina Rossetti

Remember me when I am gone away,
Gone far away into the silent land,
When you can no more hold me by the hand,
Nor I half turn to go yet turning stay.
Remember me when no more day by day
You tell me of our future that you planned:
Only remember me; you understand
It will be late to counsel then or pray.
Yet if you should forget me for a while
And afterwards remember, do not grieve:
For if the darkness and corruption leave
A vestige of the thoughts that once I had,
Better by far you should forget and smile
Then that you should remember and be sad.

5. What is the main idea of the passage?

....

....

....

....

6. What is the theme of the passage?

....

....

....

....

7. Identify an instance of figurative language used in the passage and describe its meaning.

....

....

....

8. Identify another instance of figurative language used in the passage and describe its meaning.

....

....

....

Let's get some fitness in! Go to page 237 to try some fitness activities.

What are figures of speech?

Figures of speech are words and phrases that are not to be taken literally. Figures of speech make a text expressive. It is another name for figurative language.

Key Terms:

Euphemism - a polite or indirect way to say something. It is used in order to make unpleasant language sound more pleasant

- Example - The company laid off many people.
- Explanation - This is an example of a euphemism. Saying the employees got "laid off" is a more pleasant way to say they "got fired."

Hyperbole - an extreme exaggeration that is not meant to be taken literally

- Example - I am so hungry, I could eat a horse.
- Explanation - Obviously someone is not going to eat an entire horse, so the sentence above is an example of hyperbole because it is exaggerating how hungry the narrator is to an unrealistic extent.

Oxymoron - joining two opposite words together

- Example - I am going to order jumbo shrimp for dinner.
- Explanation - This is an example of an oxymoron because the words "jumbo" and "shrimp" are opposites. The word "jumbo" means large, but the word "shrimp" means small.

Paradox - at first glance, it is a phrase that looks like it is contradictory or wrong, but after looking closer, it is actually the truth

- Example - Always expect the unexpected.
- Explanation - At first glance, the sentence above may look incorrect; however, after reading it again, you can tell it is the truth. It is important to expect that things will always be changing.

Allusion - an indirect reference to literature, history, or something else well known

- Example - My baby sister's nose kept growing when she was telling my mom about the broken window.
- Explanation - The above sentence is an allusion or reference to the children's story "Pinocchio." The character's wooden nose would grow every time he lied. This allusion explains that the narrator's baby sister was lying to her mother.

Pun - a play on words; the words used are alike or very similar but have different meanings

- Example - I was on pins and needles at the sewing store when the owner showed me the new items on display.
- Explanation - The above sentence is an example of a pun because "pins" and "needles" are also present at sewing stores.

Read the following sentences. Identify the figure of speech used and explain its meaning.

1. After running the marathon in the heat, I drank so much water that I thought I was going to explode.

What is this figure of speech? ....

What does this figure of speech mean?

....

2. I want to read the new classic that was just released at the bookstore. It was written by my favorite author.

What is this figure of speech? ....

What does this figure of speech mean?

....

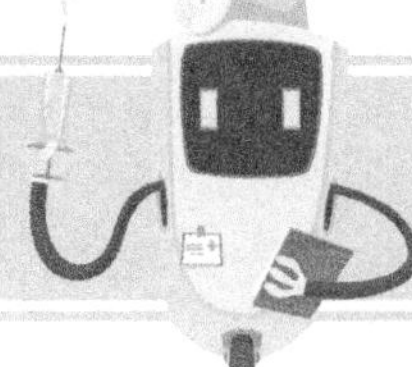

3. The lecture in math class was so terrible that I thought I was going to die of boredom.

What is this figure of speech? ....

What does this figure of speech mean?

....

4. The veterinarian told Fluffy's owner that her dog was well fed in regard to her weight.

What is this figure of speech? ....

What does this figure of speech mean?

....

5. My father told me my high school graduation was the beginning of the end.

What is this figure of speech? ....

What does this figure of speech mean?

....

6. My grandmother told my grandfather he was acting like Scrooge and needed to loosen up and have fun at the zoo.

What is this figure of speech? ....

What does this figure of speech mean?

....

7. My uncle has a farm and decided to purchase a donkey. He thought he would get a kick out of it.

What is this figure of speech? ....

What does this figure of speech mean?

....

8. My father called me Romeo, but then he gave me twenty dollars to take my new girlfriend out on a date.

What is this figure of speech? ....

What does this figure of speech mean?

....

9. In the winter, my puppy wears a little sweater when he goes outside. In the summer, my puppy pants.

What is this figure of speech? ....

What does this figure of speech mean?

....

10. If I know one thing, it is that I know nothing.

What is this figure of speech? ....

What does this figure of speech mean?

....

11. The last day of school was bittersweet.

What is this figure of speech? ....

What does this figure of speech mean?

....

12. My goldfish, Mila, passed away this morning while I was at school.

What is this figure of speech? ....

What does this figure of speech mean?

....

Let's get some fitness in! Go to page 237 to try some fitness activities.

Integers with Negative Exponents and their Equivalent Fractions

Review:

An integer with a negative exponent can also be expressed as a positive fraction with one as the numerator and the integer with the positive form of the exponent as the denominator. These are equivalent expressions.

Question 1:

What is the equivalent expression for 6^{-6}?

A. $\frac{6}{-6}$

B. $\frac{1}{6^{-6}}$

C. $\frac{1}{36}$

D. $\frac{1}{6^{6}}$

Question 3:

What is the equivalent expression for $\frac{1}{3^4}$?

A. $\frac{1}{3^{-4}}$

B. 3^{-4}

C. 3^{4}

D. $\frac{1}{12}$

Question 2:

What is the equivalent expression for 2^{-10}?

A. $\frac{1}{2^{10}}$

B. $\frac{1}{-20}$

C. $\frac{1}{2^{-10}}$

D. $\frac{-2}{10}$

Question 4:

What is the equivalent expression for $\frac{1}{10^8}$?

A. 10^{-8}

B. $\frac{1}{80}$

C. $\frac{1}{10^{-8}}$

D. 10^{8}

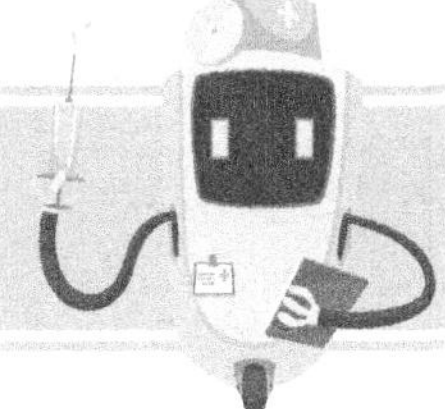

Question 5:

Water Fire is a master at transforming integers with negative exponents to their equivalent fraction. What is the fraction she creates for 15^{-6}?

Question 7:

Just as she knows a wave of water from a blaze of fire, Water Fire is certain $\frac{1}{7^6}$ has an equivalent expression in the form of an integer with an exponent. What is it?

Question 6:

While surfing the ocean waves, Water Fire transforms 12^{-12} to its equivalent fraction. What is it?

Question 8:

As Water Fire lights a bonfire on the beach, she clearly sees the equivalent expression for $\frac{1}{9^{27}}$. What is it?

Let's get some fitness in! Go to page 237 to try some fitness activities.

Perfect Squares and Perfect Square Roots

Review:

A perfect square number is a whole number that is the product of another whole number multiplied by itself. A perfect square number is always positive. The number being multiplied is called the perfect square root and can be positive or negative.

Question 1:

Which whole number is a perfect square number?

A. 10
B. 100
C. 1,000
D. None of the above

Question 2:

Which whole number is a perfect square number?

A. 81
B. -81
C. 25
D. A and C

Question 3:

What is the square root of 25?

A. 5
B. -5
C. 12.5
D. A and B

Question 4:

What is the square root of 64?

A. 8
B. 32
C. -8
D. A and C

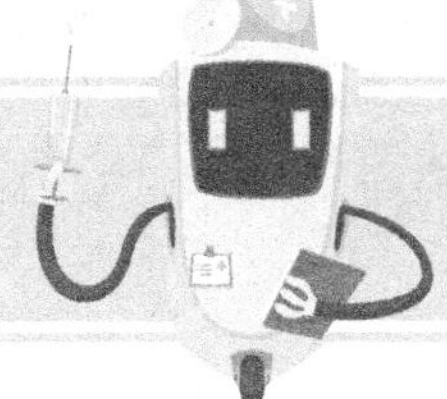

Question 5:

Water Fire understands the perfection of the tides. She also understands perfect square roots. She believes **225** has a perfect square root. Is she correct? Why or why not?

Question 6:

While transforming a droplet of water into a fiery spark, Water Fire wonders if 1 million is a perfect square. Is it? Why or why not?

Question 7:

Water Fire draws a **75** in the sand. Is this number a perfect square? Why or why not?

Question 8:

Water Fire calculates the square root of **2,500** while watching the sun set like a giant fireball over the ocean horizon. What is it?

Let's get some fitness in! Go to page 237 to try some fitness activities.

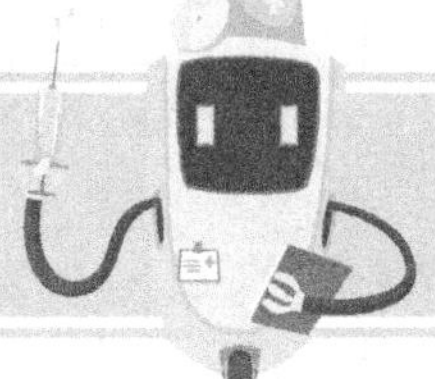

Perfect Cubes and Perfect Cube Roots

Review:

A perfect cube is a number that is the product of a number multiplied by itself and by itself again. It can be positive or negative. The number being multiplied is called the cube root and can be either positive or negative.

Question 1:

Which number is a perfect cube?

A. 10
B. 100
C. 1,000
D. None of the above

Question 2:

Which number is a perfect cube?

A. 25
B. 125
C. 150
D. None of the above

Question 3:

What is the cube root of 27?

A. 3
B. -3
C. 9
D. None of the above

Question 4:

What is the cube root of 64?

A. 8
B. 4
C. Both A and B
D. None of the above

Question 5:

While transforming a batch of ocean water into freshwater, Water Fire contemplates perfect cubes. She is certain -216 has a perfect cube root. Is she correct? Why or why not?

Question 6:

Water Fire loves to paddleboard in the ocean and thinks about cube roots as she paddles. Is 12 a perfect cube? Why or why not?

Question 7:

Drawing in the sand, Water Fire works out the cube root of -343. What is it?

Question 8:

Water Fire bobs calmly in the waves and realizes 1 million not only has a perfect square root but also a perfect cube root. What is the cube root?

Let's get some fitness in! Go to page 237 to try some fitness activities.

The Point of Freezing
SCIENCE EXPERIMENT

Key Terms:

Freezing Point Depression -adding a substance to lower the freezing point of a liquid

Solution - a liquid mixture

Solvent - an element that dissolves in a solution

Did you know that salt lowers the freezing point of water? Think about it! What happens on snowy and icy days? The city sends out salt trucks to distribute salt on the road. This makes the ice melt so it does not cause accidents for drivers.

Water freezes at zero degrees Celsius, but adding salt to a solution makes it freeze at a much lower temperature. This process is not just used in making roads safer, but in other everyday uses as well. Making the freezing temperature lower is known as freezing point depression. This scientific phenomenon does not just happen with water, but with other liquids or solutions as well. In today's experiment, you will add a solvent to your solution and test freezing point depression.

Materials Needed

1 tablespoon of sugar	$\frac{1}{2}$ cup of milk	$\frac{1}{4}$ teaspoon of vanilla extract
Sealed baggie (quart)	4 cups of crushed ice	Sealed baggie (gallon)
1 cup of salt	Towel or gloves	

Procedure

1. Add one tablespoon of sugar, a half cup of milk, and one fourth teaspoon of vanilla extract to a quart-size sealed baggie.
2. Seal the baggie and shake it to mix up the ingredients.
3. Add four cups of crushed ice to a gallon-size baggie, as well as a cup of salt.
4. Place the small baggie (quart-size) into the large baggie (gallon-size), and seal the large baggie.
5. Place a towel around the large baggie or put on gloves.
6. Shake the baggie for five minutes.

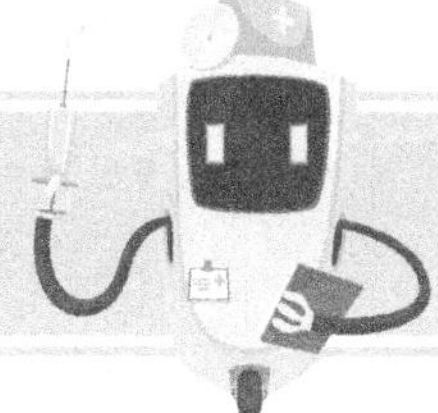

Questions

1. What product did you make in this freezing point depression experiment? Were you surprised?

2. In this experiment, what was the solvent and what was the solution?

3. What would happen if the cup of salt was not added to the baggie of ice? How would this have changed the experiment?

4. Explain the experiment using the three key terms above (freezing point depression, solution, and solvent).

5. What other everyday things occur due to freezing point depression?

Let's get some fitness in! Go to page 237 to try some fitness activities.

WEEK 3
DAY 7

MAZE

Directions: Connect the planets and the spaceships. Determine the home planet of each of them.

1 ____ 2 ____ 3 ____ 4 ____ 5 ____

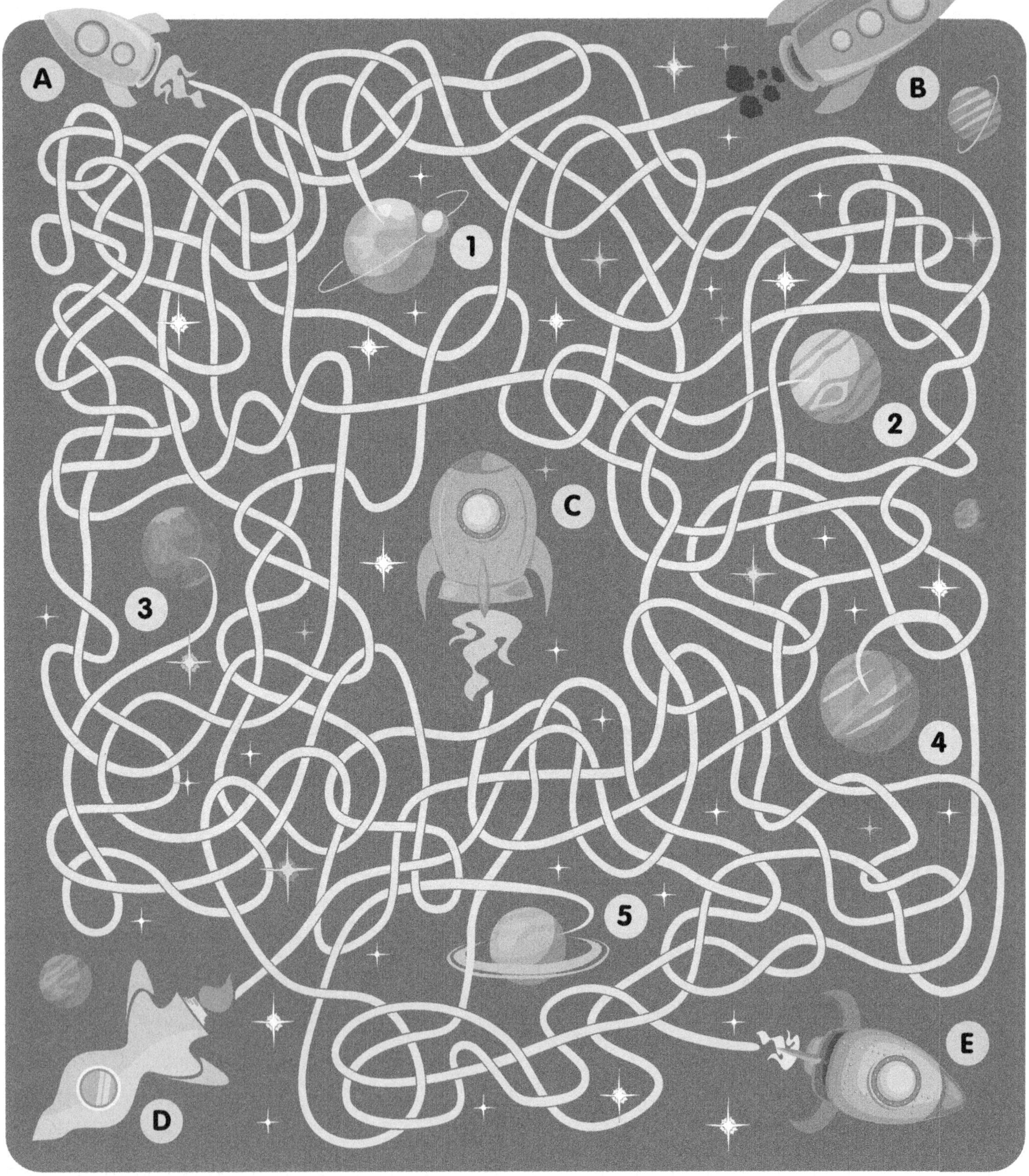

A B C D E

1 2 3 4 5

WEEK 4

GRADE 8-9

You're a rockstar!
In Week 4, we will look into compare / contrast and cause / effect text structures. Math this week is all about scientific notations!

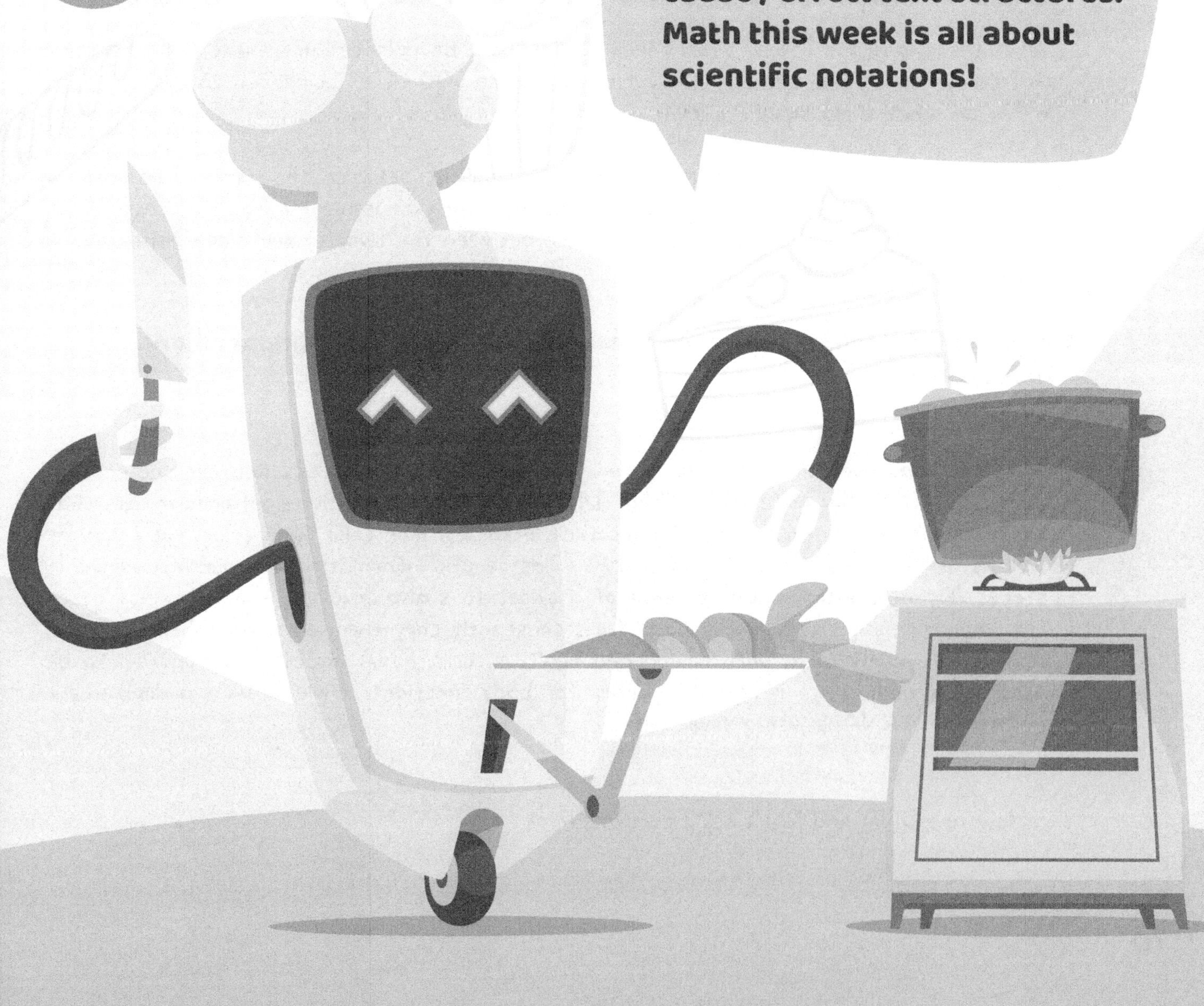

Part 1: Compare and Contrast

What is the compare and contrast text structure?

A text that uses a compare and contrast organizational structure compares or highlights similarities and differences. When you are trying to identify a certain text structure, which will help you comprehend the text and find the purpose, you need to look for certain keywords that will show you the author is comparing and contrasting. When an author uses a compare and contrast structure, you can look for some keywords that stand out. When comparing, authors generally use words like "same," "both," and "similarly." When an author is contrasting, or showing differences, some keywords include "however," "on the other hand," and "differences."

Here is an example of a sentence that contains compare and contrast structure:

While Disneyland and Disney World seem similar, both being large amusement parks centered on Mickey Mouse, Disneyland in California is a mere 500 acres compared to Disney World in Florida which is 43 square miles.

In the above example, readers can see the comparisons between the parks. The parks in Disneyland and Disney World are similar. They are both amusement parks and have Disney characters like Mickey Mouse. The difference between the two is their size; Disney World in Florida is a lot larger.

Read the paragraphs below and identify what is being compared and contrasted.

In science class, you have learned, or will soon learn, about genetics and the presence of DNA and RNA. You have learned that DNA is genetic information contained in cells that explain someone's genetics. This includes their ethnicity, hair color, skin color, eye color, as well as other pertinent information. DNA holds genetic information in a person's genes. It stores this information similar to that of a closet. It is also crucial in the remake of other cells and replenishes cells in the body. Cells constantly copy themselves and then duplicate, always leaving the body with newer and healthier cells. RNA in cells make protein to be used in the body, which is needed in order for body functions. While DNA is a double helix in structure, RNA is just a single helix.

1. How are DNA and RNA being compared?

....

2. How are DNA and RNA different?

....

....

The word revolution means a forcible overthrow of a government. The American Revolution occurred for eighteen years, starting in 1765. The American Revolution began in an attempt to fight the monarchy of England that was oppressing the colonists living in America. Even though America and England were an ocean apart, England still had control over the colonists living in America. The colonists did not like England's rule over their government and religion, nor did they like the high taxes. They did not want the king to rule them anymore. The colonists wanted to be independent from Britain and wanted to establish their own form of government. A different revolution occured two decades later in France: the French Revolution in 1789. Like America, the country of France wanted to be independent, not from another country but from old and rich aristocrats who ruled over them. People in France did not like the direction their country was headed in. They wanted to break away from the traditional ways and start anew.

3. How are the American Revolution and the French Revolution similar?

....

....

4. How are the American Revolution and the French Revolution different?

....

....

Part 2: Cause and Effect

What is the cause and effect text structure?

The cause and effect organizational structure is another text structure that authors use in order to explain an action that aligns with the author's purpose. This particular structure shows the audience how certain actions and events are connected. Just like the compare and contrast structure, there are keywords which help readers pick up on the cause and effect structure. Some keywords that show readers the link of cause and effect include: "because," "consequently," and "as a result."

Here is an example of a sentence that contains cause and effect structure:

In many overpopulated areas, researchers are discovering that breathing problems, like asthma and emphysema, are results of poor air quality.

In the above example, the cause is poor air quality in over populated areas like large cities. The effect, or what happens due to the poor air quality, is that people have developed breathing issues.

Read the paragraphs below and identify the cause and the effect.

Individuals most likely have heard of Neil Armstrong, Buzz Aldrin, and John Glenn. These are famous astronauts documented in history books, but have you ever heard of Sally Ride? When Sally Ride was a doctoral student at Stanford University, she became interested in physics and the NASA organization. At that time, NASA was in need of scientists for their next launch. Sally Ride applied to work for NASA, even though women were seldom trained as astronauts during that time. However, despite gender disparity, Ride was accepted into the NASA program, and after sufficient training, she became the first American woman sent into space. Ride was on the Challenger Shuttle for two missions. Her third mission was canceled as a result of the tragic explosion of the Challenger a few years later. Sally Ride then decided to leave NASA in order to help others. Ride developed a school in order to inspire young girls in the areas of math and science.

5. What was the cause in the passage?

....

....

6. What is the effect based on the cause in the passage?

....

....

Diego Armando Maradona grew up in Argentina in 1961. He grew up very poor and often did not have any food to eat. He lived in a one-room house with his seven brothers and sisters. He didn't go to school and rarely wore shoes because his family could not afford them. When he grew up, he saw children his age playing soccer near him, but he did not have enough money to buy a soccer ball. He decided to make his own ball by stuffing rags in a bag and tying it up in a ball. Not having a proper ball did not stop Maradona, as he grew up to be a strong soccer player. People noticed his skills when he was fifteen, and they signed him with a professional team. At twenty one, he joined a pro soccer team in Barcelona.

7. What was the cause in the passage?

....

....

8. What is the effect based on the cause in the passage?

....

....

Let's get some fitness in! Go to page 237 to try some fitness activities.

Part 1: Choosing A Topic

In English class, does your teacher ever tell you it is time to write, but then you spend a long time figuring out what in the world you are going to write about? It is important that when you pick a topic, it is not too broad or general. It is also equally important to make sure you do not choose a topic that is very specific because then you may not have a lot to write about.

Let's look at an example:

What if your English teacher told you to write about a topic you enjoy. What if you picked football? Would football be a good topic? Probably not. Football is actually too broad because there is just so much to talk about regarding football.

Read the following topics and explain which one is the best to write about and why?

Topic 1 - The best soccer players in the world

Topic 2 - Soccer player Cristiano Ronaldo

Topic 3 - The World Cup

What is the best topic to write about and why are the others not strong topics?

....

....

Name some topics that would be too broad to write about and explain why.

- ....
- ....
- ....
- ....

Name some topics that would be too specific to write about and explain why.

- ....
- ....
- ....

Name some topics that would be ideal and make strong essay topics.

- ....
- ....
- ....

Choose an historical figure and brainstorm three possible topics for an essay about them.

Topic 1 - ....

Topic 2 - ....

Topic 3 - ....

Part 2: Thesis Statement

The thesis statement is the most important sentence in an essay. It explains what you will be writing about and what you will be proving in the essay. The thesis statement is the main claim of your essay. For example, consider this thesis statement: Who was one of the most influential judges on the Supreme Court? Does this thesis statement work? What about this one: Ruth Bader Ginsberg was a judge on the Supreme Court. Both of these thesis statements are weak. Neither one of them are clear or explain what points the writer will be discussing in the essay. Remember that a thesis statement does not mention the writer's personal opinion, and it should not be in the form of a question or a statement of fact.

1. Write a thesis statement for an essay about Ruth Bader Ginsberg that reflects the points you will discuss in an essay.

....

....

2. Investigate your historical figure, and write three clear thesis statements about them.

Thesis Statement # 1 - ....

Thesis Sentence # 2 - ....

Thesis Sentence # 3 - ....

Part 3: Topic Sentences

Remember your writing needs to be well organized. In the main body of your essay, each paragraph will need to have a topic sentence, evidence, analysis, and a concluding sentence. The body paragraphs should start with a topic sentence. Topic sentences help organize your writing and allow the readers to know what each particular paragraph will be about. The topic sentences should also align with the essay's topic and the essay's thesis statement. Similar to picking your topic, you want to make sure your topic sentences are not too general or too specific. You also need to make sure that the topic sentences align with what you write in the body paragraph.

For example, let's pretend you are writing an essay about the hit musical *Hamilton*.

Read this topic sentence for an essay about *Hamilton*: Lin-Manuel Miranda wrote the musical *Hamilton* based on a biography about Alexander Hamilton's life. What information would fit in this particular paragraph? Would you write about how the musical was developed? Would you write about what awards the musical received? No. Neither of those questions fit or align with the topic sentence. If you included this information in the body paragraph, the paragraph would contain way too much information, and it would not align with the topic sentence.

Instead, the paragraph should contain information about how Lin-Manuel Miranda came up with the idea for the musical and how the biography about Alexander Hamilton inspired him to write the musical.

Go back to the thesis statements you wrote about your historical figure. Pick one of the thesis statements and write it on the line below.

....

Based on the thesis statement above, what three sections or paragraphs do you think you would include in your essay?

Paragraph # 1 - ....

Paragraph # 2 - ....

Paragraph # 3 - ....

In the three sections above, what would your topic sentences be for each paragraph? Write the topic sentences on the lines below.

Topic Sentence # 1 - ....

Topic Sentence # 2 - ....

Topic Sentence # 3 - ....

Let's get some fitness in! Go to page 237 to try some fitness activities.

Scientific Notation and Positive Exponents

Review:

Scientific notation is used to represent very large numbers. Large numbers are in the form of a positive number between 1 and 10 multiplied by a power of 10 with a positive exponent. A positive exponent represents how many times the decimal is moved to the right.

Question 1:

Which scientific notation represents 6,000,000,000?

A. 6×10^9

B. 6×10^{10}

C. 6×10^8

D. 60×10^8

Question 2:

Which scientific notation represents 785,000,000?

A. 785×10^6

B. 78.5×10^7

C. 7.85×10^8

D. $.785 \times 10^9$

Question 3:

What is the expanded form of 9×10^{12}?

A. 900,000,000,000

B. 9,000,000,000,000

C. 9,000,000,000

D. 90,000,000,000,000

Question 4:

What is the expanded form of 5.6×10^8?

A. 560,000

B. 56,000,000

C. 560,000,000

D. 5,600,000

Question 5:

Thunder Warrior's motto is bigger is always better. He creates a tidal wave 8500 meters high. What is this number in scientific notation?

Question 7:

It's time for Thunder Warrior to show off the thing he does best: thunder. He creates a rumble which can be heard 8.5 x 10^4 miles away. How many miles is this in expanded form?

Question 6:

Thunder Warrior loves a good tsunami in the middle of the ocean. He creates a new class of storm traveling 56,700 miles per hour. What is this number in scientific notation?

Question 8:

Thunder Warrior knows lightning strikes at the speed of light, approximately 1.86 x 10^5 miles per second. What is this number in expanded form?

Let's get some fitness in! Go to page 237 to try some fitness activities.

Scientific Notation and Negative Exponents

Review:

Scientific notation is also used to represent very small numbers. A small number is in the form of a positive number between 1 and 10 multiplied by a power of 10 with a negative exponent. The negative exponent represents how many times the decimal is moved to the left.

Question 1:

Which scientific notation represents 0.000009?

A. 9×10^{6}
B. 9×10^{-6}
C. 90×10^{-7}
D. None of the above

Question 2:

Which scientific notation represents 0.012?

A. 12×10^{-3}
B. 12×10^{3}
C. 1.2×10^{-2}
D. None of the above

Question 3:

What is the expanded form of 5×10^{-6}?

A. 0.000005
B. 0.00005
C. 5,000,000
D. None of the above

Question 4:

What is the expanded form of 2×10^{-2}?

A. 0.2
B. 0.02
C. 0.002
D. None of the above

Question 5:

Thunder Warrior's sidekick, Hurricane, prefers small numbers. For fun, he spins sand particles that are 0.000097 millimeters in diameter. What is this number in scientific notation?

Question 6:

Hurricane creates a miniature whirlwind on the beach in 0.0036 seconds. What is this number in scientific notation?

Question 7:

When Hurricane is feeling especially tired, he spins around at a lazy 5×10^{-1} kilometers per hour. What is this speed in expanded form?

Question 8:

For Thunder Warrior's birthday, Hurricane whips up a batch of saltwater taffy in 9.99×10^{-2} seconds. What is this time in expanded form?

Let's get some fitness in! Go to page 237 to try some fitness activities.

Performing Operations with Scientific Notation

Review:

To perform any operation between two numbers in scientific notation, the exponents must be the same. If they are the same, then the lead numbers are added or subtracted and the 10 with its exponent is added onto the answer. If they are not the same, one of the numbers must first be rewritten so the exponents match. The final answer must always be in proper scientific notation.

Question 1:

Solve $6 \times 10^6 + 5 \times 10^6$

A. 11×10^6
B. 30×10^6
C. 30×10^{12}
D. 11×10^{12}

Question 2:

Solve $12 \times 10^3 - 6 \times 10^3$

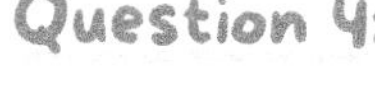

A. 6×10^6
B. 6×10^3
C. 18×10^3
D. 72×10^3

Question 3:

Solve $4 \times 10^2 + 4 \times 10^3$

A. 4.4×10^3
B. 8×10^6
C. 8×10^2
D. 8×10^5

Question 4:

Solve $5 \times 10^5 + 5 \times 10^4$

A. 10×10^1
B. 10×10^9
C. 5×10^5
D. 5.5×10^5

Question 5:

Thunder Warrior is adding up the forces of two lightning bolts. What is the sum of $9 \times 10^8 + 10 \times 10^8$?

Question 6:

Thunder Warrior and his sidekick Hurricane are comparing their wind speeds. Solve $5 \times 10^7 - 4 \times 10^7$.

Question 7:

Thunder Warrior likes to throw his sound waves around. He knows their power to create ocean storms equals $1.2 \times 10^6 + 8 \times 10^5$. What is this sum in proper scientific notation?

Question 8:

Hurricane rained out a beach barbecue for 3×10^5 minutes, paused, and then rained for an additional 4×10^6 minutes. How many minutes did it rain on the beach? Write this sum in proper scientific notation.

Let's get some fitness in! Go to page 237 to try some fitness activities.

Can Mountains Grow?

Mount Everest is the world's tallest mountain, blanketing the countries of Nepal, Tibet, and China. Many individuals flock to Mount Everest in order to climb it, even though this is no easy feat. There are many physical dangers to climbing this mountain, including the hazardous altitude and the frequent avalanches. Because Mount Everest is so high, climbing it threatens our bodies. At over 19,000 feet, it is impossible for humans to get enough oxygen from the air. A lack of oxygen weakens the human body and can cause serious respiratory distress as well as heart conditions. Unfortunately, climbers have trouble seeking help so high in elevation. Many avid climbers bring oxygen with them, but oxygen is heavy to carry, expensive, and can be unreliable if the tanks malfunction. There's also the possibility it may run out.

How do you reach the top of Mount Everest? Just go up, right? Well yes, but there are over fifteen different ways to reach the top. Two routes, one in Tibet and one in Nepal, are the ones most commonly traversed. Even these routes take about five days round trip. In the last two years, a total of 295 people have died on their journey to reach the peak. In 1953, Edmund Hillary was the first person to climb to the highest point of Mount Everest at 29,032 feet. In 2019, Rita Sherpa climbed the same peak to 29,035 feet. Sherpa climbed higher than Hillary, but they were at the top of the same peak. How? Well, because Mount Everest has grown since 1953!

That's right. Mount Everest is growing, but how? Well, it is a young mountain. Even though Mount Everest is over fifty million years old, it is in a young mountain range formed by the collision of two tectonic plates creating the tallest mountain in the world. But it is constantly getting taller. In late 2020, China measured the height of the highest point of Mount Everest and reported it continues to grow higher. The tallest peak has increased by three feet since the 1950s. This is not a fluke either, since this measurement was checked and rechecked to ensure accuracy. Scientists did not use a measuring stick to measure this mountain. Instead, they used survey leveling data, trigonometry, and satellites to determine its height. Geologists' explanation for the growth of Mount Everest is due to the shifting tectonic plates. The plates push Earth's outer layer up and out, making the mountain taller. The opposite can happen as well. Earthquakes make many mountains sink, but this has not happened to the ever-growing Mount Everest.

1. What other physical dangers might harm climbers of Mount Everest?

2. What physically happens to the human body when oxygen is depleted? Why is there not enough oxygen on top of Mount Everest?

3. Despite the mountain growing, do you think it was harder to climb Mount Everest in the 1950's compared to now? Explain.

4. How do earthquakes make mountains sink?

5. Look on the internet and research which mountains have increased their height due to the shifting of tectonic plates and which mountains have decreased in height due to earthquakes.

MAZE

Directions:

After the concert, the guitar cables were tangled. Help the musicians by connecting the numbered guitars with the lettered speakers.

1 - ☐

2 - ☐

3 - ☐

4 - ☐

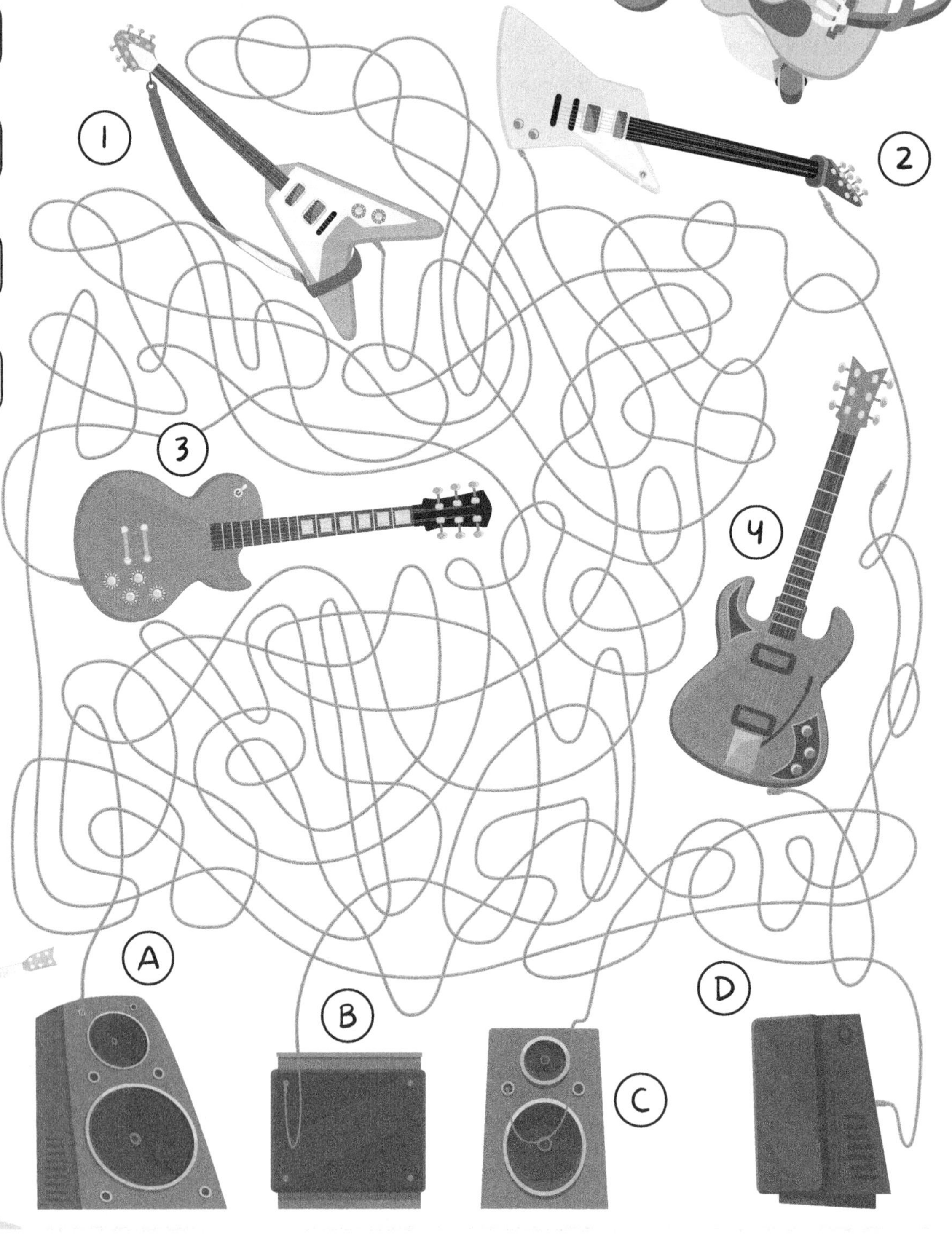

1 2 3 4

A B C D

WEEK 5

GRADE 8-9

Week 5 is all about claims and evidence when it comes to writing. After that, you'll work on linear equations and a super fun experiment on soil erosion!

Part 1: Problem and Solution

Similar to the compare and contrast and the cause and effect structures, the problem and solution structure describes an author's organization. Writers use this text structure to show a particular problem and then give a particular solution to that specific problem.

Like the other organizational structures, there are keywords which alert readers to this particular structure. Some words to look for in a problem and solution structure are "question," "answer," and "solve."

Here is an example of this writing structure:

People need to make sure they do not flush wet wipes down the toilet, or it will overflow like it did last summer.

The problem in the above example is that the toilet overflowed last summer. The solution is to make sure people do not flush any more wet wipes down the toilet.

Read the following paragraphs and explain the problem and solution within the paragraph.

During World War II, Winston Churchill was the prime minister of the United Kingdom. Like many American wartime heroes, such as George Washington and Charles Linbergh, Churchill was a war hero in his time. He was a leader many people looked up too; however, when he was growing up, no one looked up to him. As a child, Churchill had a speech impediment, and many people made fun of the way he talked. Churchill did not let that stop him, and he worked hard to overcome his speech difficulties. Also, Churchill had a lot of trouble in school. He struggled in his studies, and many of his peers made fun of him. Again, Churchill did not let that stop him. He worked hard and pushed ahead. Eventually, he enrolled in a prestigious military college.

1. What is the problem in this passage?

....

....

2. How was the problem solved?

....

Joelle was an avid gardener. She tried everything to keep her plants strong and healthy in order to win blue ribbons at the county fairs. She made sure her tomatoes did not rot. In order to prevent rotting, she added crushed eggshells to the soil to improve it. The eggshells prevented soil rot due to the presence of calcium, which killed fungal diseases. Joelle also added baking soda and water in a spray bottle. She would spray her plants with the mixture. This kept aphids, little white insects, away. She also bought ladybugs, which feasted on aphids.

3. What is the problem in this passage?

....

....

4. How was the problem solved?

....

....

Part 2: Structure Review

Now that you know the different structures authors use, let's review them in your own words.

1. Cause and Effect - ....

2. Compare and Contrast - ....

3. Problem and Solution - ....

Read the following paragraphs. Determine their structure and explain why it is that particular structure.

The reason why some people put nets up where they live, wear long sleeves in hot weather, and spray mosquito repellent constantly is because mosquitoes can spread terrible diseases with their bite. Mosquitos carry certain diseases like Malaria and the Zika Virus. These are serious diseases that harm and can even kill humans. Though the Zika Virus is less harmful than Malaria, it is quite dangerous to pregnant women.

1. What is this type of organizational structure?

....

2. What are the keywords that show you this particular structure?

....

3. How do you know it is this structure? Explain.

....

Have you ever wondered why some people have light hair and others have dark hair? Hair color is genetic, but the color of a person's hair is dependent upon how much melanin is in a person's hair. In melanin, there are dark and light pigments. When the hair grows out of a hair follicle, the pigments cells in that follicle determine its color. When people get older, melanin decreases. This is the reason people end up having gray hair as they age.

4. What is this type of organizational structure?

....

5. What are the keywords that show you this particular structure?

....

6. How do you know it is this type of organizational structure? Explain.

....

....

Students practice fire drills at school and administrators urge the importance of taking them seriously for safety. But why are there not fire drills for other locations? Even though fire drills are very common in school environments, fire department specialists think families should be prepared for fire evacuations at home, as well as school. Experts say it is important to discuss fire safety with your family in order to know what each person in the household needs to do in a fire and how to find the best exit depending on your location in the house. Fire safety specialists explain the importance of knowing all exits in a house. They also urge the importance of locating and maintaining the smoke detectors, carbon dioxide detectors, and fire extinguishers.

7. What is this type of organizational structure?

....

8. What are the keywords that show you this particular structure?

....

9. How do you know? Explain.

....

10. Write a couple of sentences that show a problem and solution organizational text structure.
(Please use a seperate piece of paper to answer questions 10 - 13.)

....

11. Write a couple of sentences that show a cause and effect organizational text structure.

....

12. Write a couple of sentences that show a compare and contrast organizational text structure.

....

13. Write a couple of sentences that show a chronological or sequential text structure.

....

Let's get some fitness in! Go to page 237 to try some fitness activities.

Part 1: Claims and Evidence

Remember that a claim is what you are trying to prove. Writers make claims in thesis statements, as well as in the topic sentences, in order to emphasize what they will be discussing in either the entire essay or in that particular paragraph. The evidence provided after the topic sentences in the body paragraphs needs to be able to prove the claim. If the evidence does not prove the claim, then the claim will remain insufficient.

Let's practice by looking at this claim:

Smartphones are a major distraction to teenagers.
What evidence should you include in a paragraph if this was your topic sentence? What about information explaining how smartphones make it easier to cheat or talk with friends in school rather than listen to teachers? When writing, you need strong evidence to support claims to make sure they are valid and proven.

Below are some claims. Write what evidence you would need in order to explain and make the claims valid.

1. The next presidential turn is intended to improve the economy.

....

2. Schools in America should make financial literacy classes mandatory.

....

3. Learning another language is important for everyone and should be required in high school.

....

4. Cursive handwriting should be taught in elementary schools.

....

5. The national voting age should be changed to sixteen years of age.

....

The evidence that backs up claims should be written with details, facts, and quotations you receive from outside sources like a textbook or the internet.

Going back to the claims on the previous page, find two pieces of evidence for each claim. Make sure you include the website you used in order to show where you received this information. Also, remember to put the word-for-word information you found in quotes so you you are not plagiarizing someone else's work. More lessons on plagiarism and citations will be taught in future lessons.

Here is an example:

Claim: Cursive writing should be taught in school.

Evidence # 1 - "Cursive writing defines you as much as your physical features do."

Source - https://www.edutopia.org/article/what-we-lose-with-decline-cursive-tom-berger

Evidence # 2 - "Cursive writing develops motor skills and reinforces learning."

Source - https://resilienteducator.com/classroom-resources/5-reasons-cursive-writing-should-be-taught-in-school/

For the following claims, find two pieces of evidence for each and note their source.

6. Global warming needs to be addressed immediately.

Evidence # 1 - ....

Source - ....

Evidence # 2 - ....

Source - ....

7. Public schools in America should make financial literacy classes mandatory for high school students.

Evidence # 1 - ....

Source - ....

Evidence # 2 - ....

Source - ....

8. Learning another language is important and can help individuals in future careers.

Evidence # 1 - ....

Source - ....

Evidence # 2 - ....

Source - ....

9. When looking for a pet, families should consider adopting a dog or cat from an animal shelter.

Evidence # 1 - ....

Source - ....

Evidence # 2 - ....

Source - ....

10. The national voting age should be changed to sixteen years of age.

Evidence # 1 - ....

Source - ....

Evidence # 2 - ....

Source - ....

Part 2: Counterclaims

When writing, it is important to think about the other side of your argument. A counterclaim or counterargument is an argument against your claim. Addressing the opinions of the other side will demonstrate you considered the other side and how other people feel about the argument or issue. It will also strengthen your claim because you can include specific arguments against the counterclaim.

Here is an example:

If a claim is that financial literacy courses should be taught in school, the counterclaim or counterargument is that they should not be taught for a certain reason. That reason could be because high school students are already bombarded with enough required courses they have to take.

Think about the following claims. Write a counterclaim or counterargument to that claim and a reason behind the counterclaim.

1. The death penalty should be banned in all fifty states.

   Counterclaim - ....

   Reason for the Counterclaim - ....

2. Drones should be banned because they violate privacy.

   Counterclaim - ....

   Reason for the Counterclaim - ....

3. Wearing a motorcycle helmet is not necessary.

   Counterclaim - ....

   Reason for the Counterclaim - ....

4. There should be a universal healthcare system.

   Counterclaim - ....

   Reason for the Counterclaim - ....

Let's get some fitness in! Go to page 237 to try some fitness activities.

FITNESS TIME

Linear Equations

Review:

A linear equation will produce coordinate points that can be graphed. Solve for the y-coordinate by plugging the given number for the x-coordinate into the equation.

Question 1:

Use the linear equation $y = 2x + 3$ to solve for y. Graph the coordinate points and draw a line between them.

x	1	2	3	4
y				

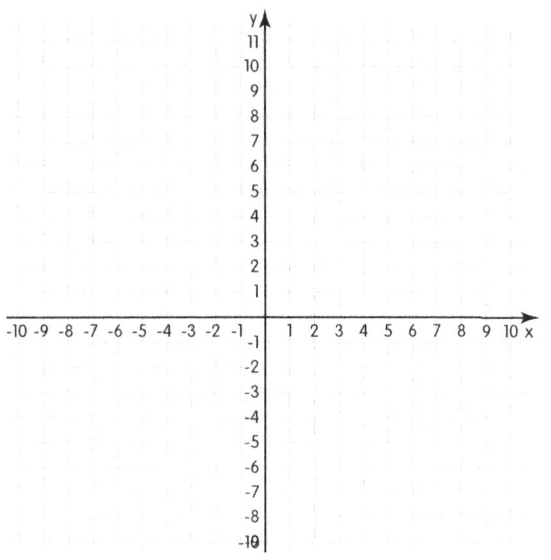

Question 2:

Use the linear equation $y = 2x - 3$ to solve for y. Graph the coordinate points and draw a line between them.

x	1	2	3	4
y				

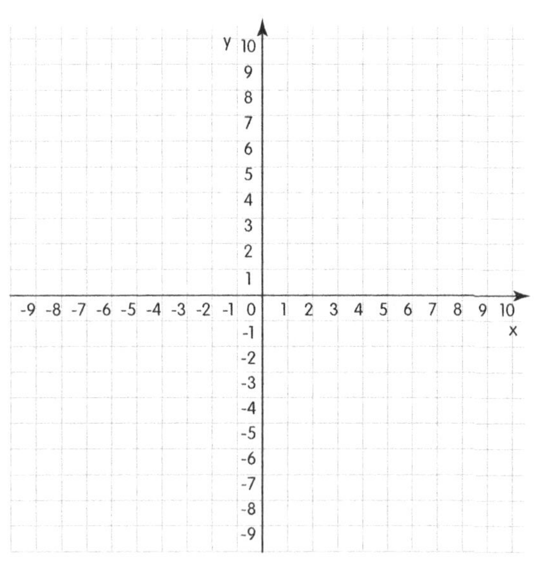

Question 3:

Use the linear equation $y = \frac{1}{2}x + 4$ to solve for y. Graph the coordinate points and draw a line between them.

x	1	2	3	4
y				

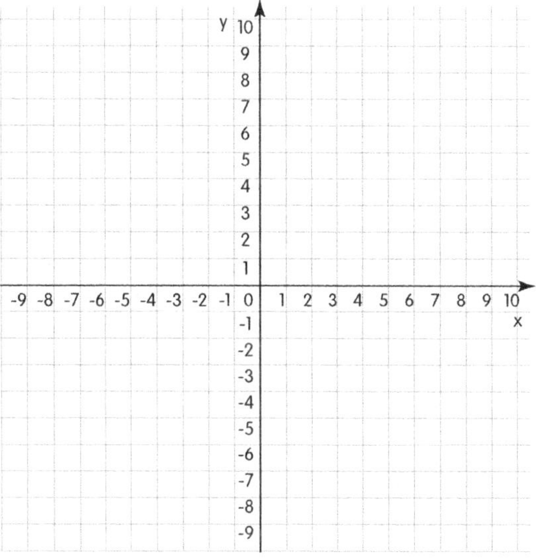

Question 4:

Use the linear equation $y = x - 5$ to solve for y. Graph the coordinate points and draw a line between them.

x	1	2	3	4
y				

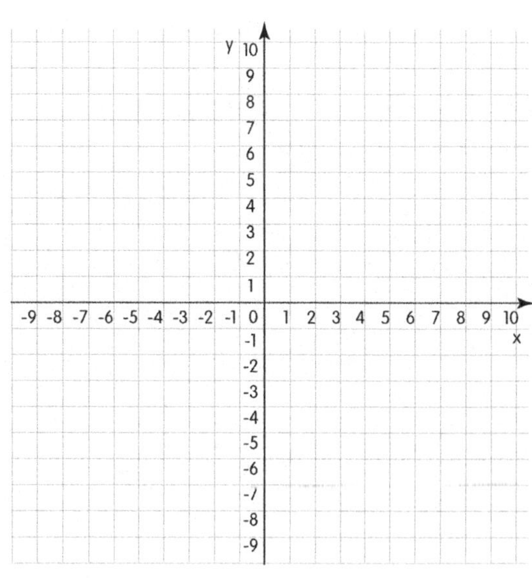

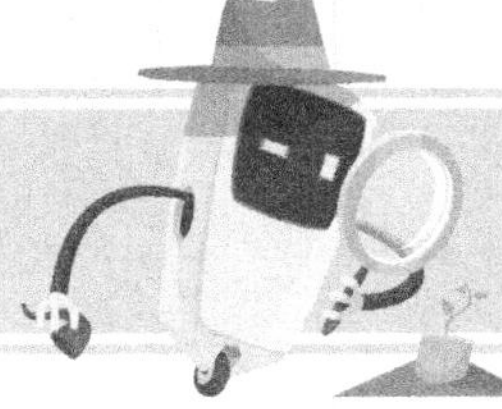

Question 5:

Rapid Ninja likes to track his speed by noting his distance in meters per second. His speed across the beach matches the linear equation: $y = 6x + 6$. Determine the coordinate points and graph them.

x	1	2	3	4
y				

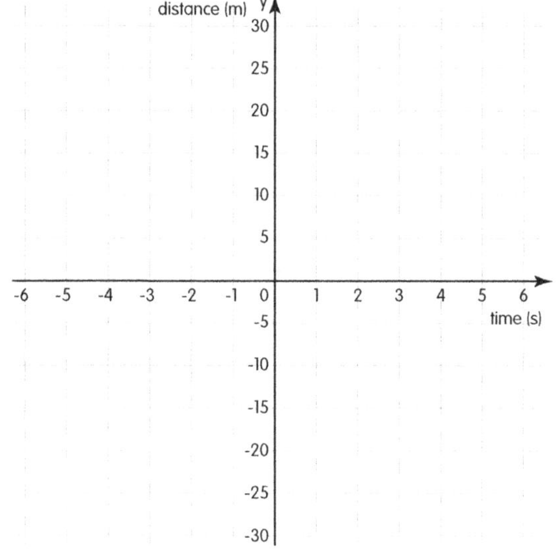

Question 6:

Rapid Ninja's speed underwater matches the linear equation: $y = 3x + 10$. Determine the coordinate points and graph them.

x	1	2	3	4
y				

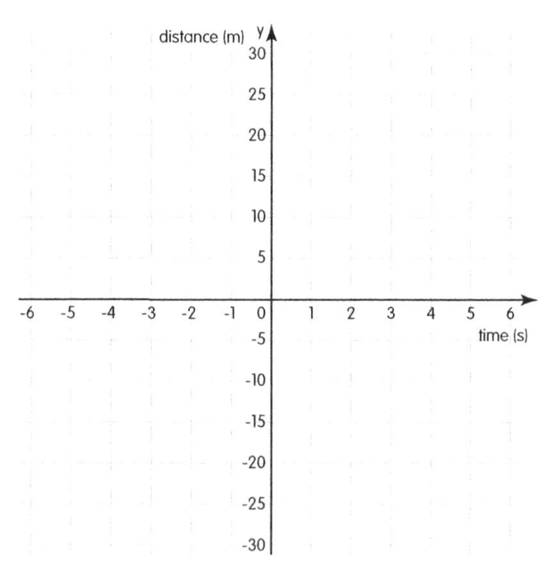

Question 7:

Rapid Ninja loves to surf. His speed through a wave tunnel matches the linear equation: $y = 5x - 2$. Determine the coordinate points and graph them.

x	1	2	3	4
y				

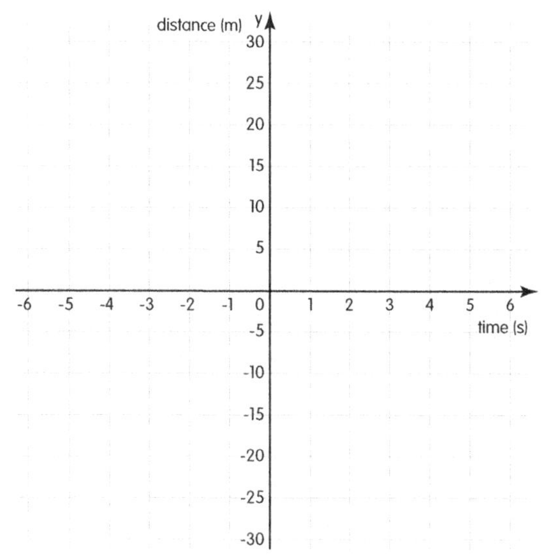

Question 8:

Rapid Ninja has just learned to paddle board. His speed across a cove matches the linear equation: $y = 4x - 1$. Determine the coordinate points and graph them.

x	1	2	3	4
y				

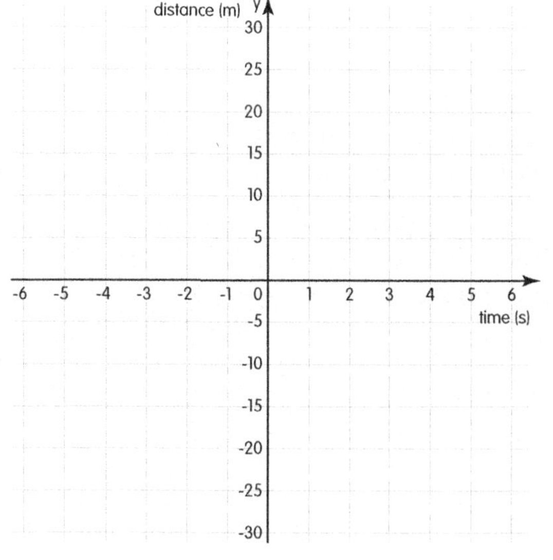

Let's get some fitness in! Go to page 237 to try some fitness activities.

Finding the Y-Intercept

Review:

The y-intercept of any line is the point where the line crosses the y-axis. To find the y-intercept, plug in "0" for x in the linear equation.

Question 1:

What is the y-intercept for the linear equation $y = 2x + 5$?

A. (0,2)
B. (0,5)
C. (0,3)
D. (5,0)

Question 3:

What is the y-intercept for the linear equation $y = \frac{1}{2}x + 5$?

A. (0,.5)
B. (.5,0)
C. (0,5)
D. (5,0)

Question 2:

What is the y-intercept for the linear equation $y = 2x - 5$?

A. (0,-5)
B. (0,5)
C. (5,0)
D. (-5,0)

Question 4:

What is the y-intercept for the linear equation $y = -3x$?

A. (0,0)
B. (0,-3)
C. (-3,0)
D. (3,3)

Question 5:

Rapid Ninja's speed following a group of dolphins matches the linear equation $y = \frac{1}{3}x + 8$. What is the y-intercept of this line?

Question 7:

The linear equation for Rapid Ninja's speed when he is chasing a giant squid is $y = 20x + 1$. What is the y-intercept of this line?

Question 6:

The linear equation for Rapid Ninja's speed when he is retreating from a hammerhead shark is $y = -5x - 10$. What is the y-intercept of this line?

Question 8:

The linear equation for Rapid Ninja's speed when he is swimming backwards from a jellyfish is $y = 8.5x - 7$. What is the y-intercept of this line?

Let's get some fitness in! Go to page 237 to try some fitness activities.

Finding the Slope of a Line

Review:

A simple linear equation is in the form of $y = mx + b$. The variable "m" represents the slope of the line. The slope of a line is represented in its simplest ratio such as $\frac{1}{2}$.

Question 1:

What is the slope in the linear equation $y = 2x + 1$?

A. $\frac{2}{1}$

B. 5

C. $\frac{1}{2}$

D. 7

Question 2:

What is the slope in the linear equation $y = \frac{1}{5}x + 4$?

A. $\frac{5}{1}$

B. .2

C. $\frac{1}{5}$

D. 1

Question 3:

What is the slope in the linear equation $y = 5x$?

A. $\frac{5}{1}$

B. 2

C. $\frac{7}{1}$

D. $\frac{1}{5}$

Question 4:

What is the slope in the linear equation $y = x$?

A. 0

B. $\frac{1}{1}$

C. 2

D. None of the above

Question 5:

Rapid Ninja chases ocean storms. His speed toward a hurricane matches the linear equation: $y = 7x$. What is the slope of this line?

Question 6:

Rapid Ninja zips away from the eye of the hurricane with a speed represented by $y = -10x$. What is the slope of this line?

Question 7:

Rapid Ninja spots a tsunami and speeds toward it. The linear equation reflecting his speed is $y = \frac{1}{5}x + 20$. What is the slope of this line?

Question 8:

Rapid Ninja escapes a tidal wave with a speed matching the linear equation $y = -\frac{5}{2}x + 6$. What is the slope of this line?

Let's get some fitness in! Go to page 237 to try some fitness activities.

Soil Erosion

SCIENCE EXPERIMENT

Key Terms:

Soil erosion -the washing or blowing away of soil

Runoff - the draining away of water

Since there is soil all around us, you may think soil conservation is unnecessary. However, a problem is plaguing soil and causing catastrophic issues. Soil erosion occurs quite frequently due to high winds and rushing water. While it does not seem like a major problem, countless fields of crops get washed away every year. This then adds more devastation since the chemicals and pesticides used on the crops find their way into the water supply. Soil erosion also causes landslides and mudslides, and this damage costs individuals and governments a lot of money. What keeps soil from doing this? The answer is plants and trees because the roots can keep the soil intact and can block out the wind as well.

Let's come up with an experiment to see how plants and trees prevent or help stop erosion.

Materials Needed

2 aluminum bread pans	2 large aluminum pans	Soil to fill the bread pans
Vegetable seeds	Water	Grass blades with their roots
A large book	Cup	

Procedure

1. Fill both bread pans with soil. This can be soil from outside or soil you buy at the store.
2. Go outside and pull up grass (ask permission first) with its roots.
3. In one of the bread pans, plant the grass and the vegetable seeds.
4. Place both pans near sunlight and water both pans every single day.
5. Repeat the watering process for a total of ten days. When watering the pans, water them gently and try to simulate rain.
6. On the eleventh day. Take both pans and put them in a larger aluminum pan. Lean the large pan against a book to simulate a hill.
7. Now that the bread pans are on an angle, water them one by one. Notice what happens to the water and the soil. Water each bread pan a total of three times from a standard cup.

Questions

1. Describe what happened to each bread pan in this experiment.

....

....

2. What are the causes and different types of soil erosion?

....

....

3. What are some economic issues relating to soil erosion?

....

....

4. Based on the results of the experiment, how do you think soil erosion could be prevented?

....

....

Let's get some fitness in! Go to page 237 to try some fitness activities.

MAZE

Directions: Robot Stan wants to fix his friend, but he is missing one last part. Help Stan get to the missing part hidden in the center of the maze.

WEEK 6

GRADE 8-9

In Week 6 we will dig in and start analyzing literature. You'll learn about rhetorical appeals and logical fallacies. For math, you will tackle slope and equations.

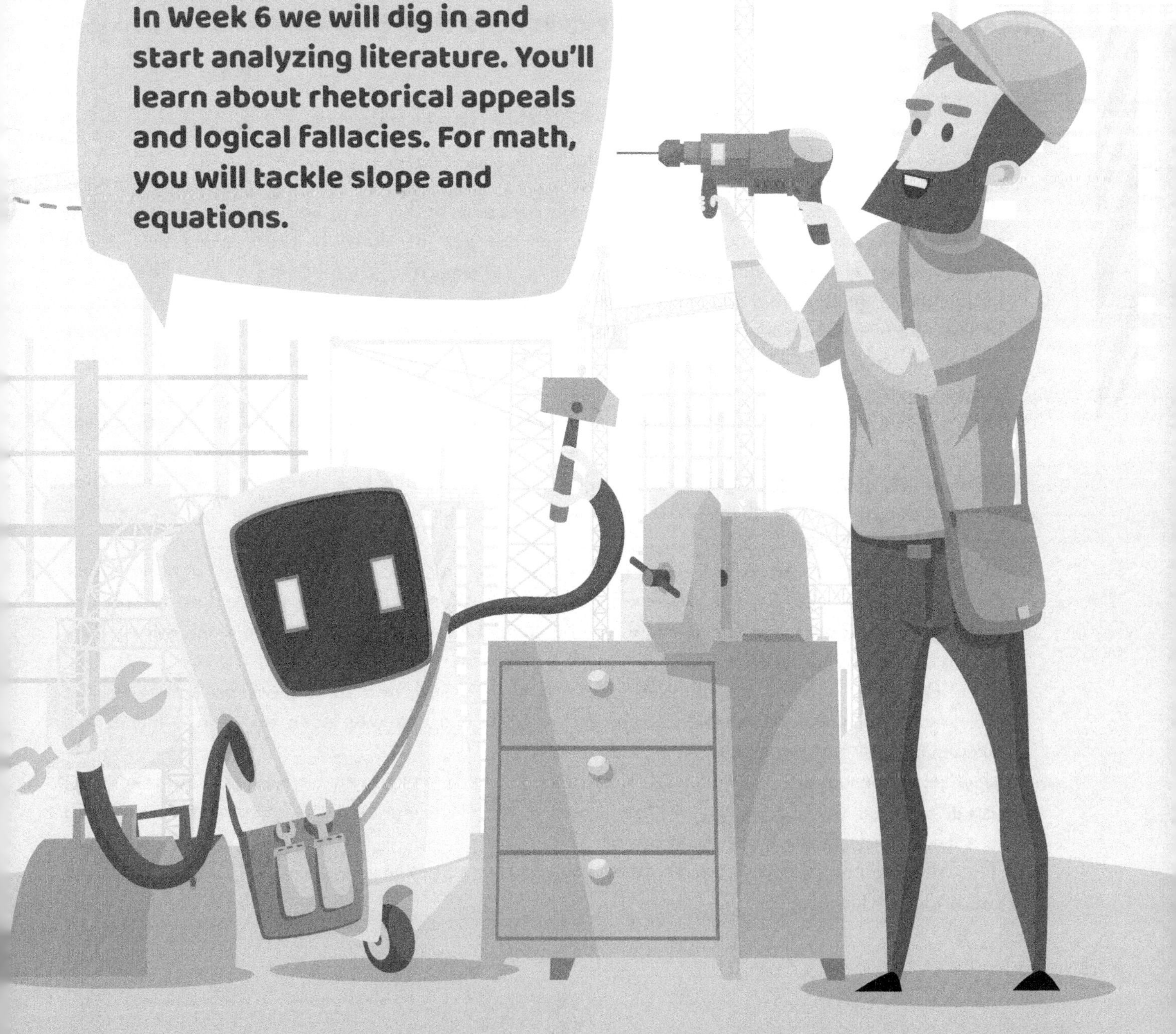

What does it mean to analyze literature?

Analyzing literature means to dig deep into a text. Analyzing literature does not mean surface level reading but reading actively and asking questions as you read. Analysis includes looking at the story elements, the actions in a text, the feelings of the characters, and the specific details in a story. Part of doing this is to make inferences or guesses about what will happen in a story. You do not have to be right, but making inferences helps you constantly think about what will happen next. Making inferences keeps you actively reading.

Read the passage and answer the questions that follow.

The Elves and the Shoemaker

(Adapted Public Domain Material)

There was once a shoemaker who, through no fault of his own, had become so poor that eventually he had only enough leather left for one pair of shoes. In the evening, he cut out the shoes, which he intended to begin the next morning, and since he had a good conscience, he lay down quietly, said his prayers, and fell asleep.

In the morning, when he had prayed, as usual, and was preparing to sit down to work, he found the pair of shoes finished on his table. He was amazed and could not understand it in the least.

He took the shoes in his hand to examine them more closely. They were so neatly sewn that not a stitch was out of place, and they were as good as the work of a master-hand.

Soon after, a purchaser came in, and he was pleased with the shoes. He paid more than the ordinary price for them. With this money, the shoemaker was able to buy leather for two pairs of shoes.

He cut the leather out in the evening, and the next day, with fresh courage, went to work. But he had no need to, for when he got up, the shoes were finished. He had customers waiting at the door, and they bought all the shoes he had. This gave him enough money that he was able to buy leather for four pairs of shoes.

Early the next morning, he found the four pairs finished, and so it went on. What he cut out in the evening was finished in the morning so that he was soon again in comfortable circumstances and became a well-to-do man.

Now it happened one evening, not long before Christmas, when he had cut out shoes as usual, that he said to his wife: "How would it be if we were to sit up tonight to see who it is that lends us such a helping hand? "

The wife agreed. She lit a candle, and they hid themselves in the corner of the room behind the hanging clothes.

At midnight came two little men who sat down at the shoemaker's table, took up the cut-out work, and began with their tiny fingers to stitch, sew, and hammer so neatly and quickly, that the shoemaker could not believe his eyes. They did not stop until everything was finished and stood complete on the table; then they ran swiftly away.

The next day the wife said: "The little men have made us rich, and we ought to show our gratitude. I will make them little coats, and you shall make them each a pair of shoes. "

The husband agreed, and that evening, when they had everything ready, they laid out the presents on the table and hid themselves to see how the little men would behave.

At midnight, they came skipping in, and were about to set to work when they found the charming little clothes and shoes.

At first, they were surprised, and then excessively delighted. With the greatest speed, they put the pretty clothes on and smoothed them down, singing: "Now we're dressed so fine and neat, why cobble more for others' feet?"

Then, they hopped and danced about, and leaped over chairs and tables as they went out the door. Henceforward, they came back no more, but the shoemaker fared well as long as he lived and had good luck in all his undertakings.

1. Why do you think the little elves stopped making shoes for the shoemaker after they received the gifts?

....

2. Why do you think the elves chose to help the shoemaker?

....

3. What would have happened if the elves never came into the shoemaker's life?

....

4. How did the shoemaker change as the story progressed?

....

5. What is the story's theme?

....

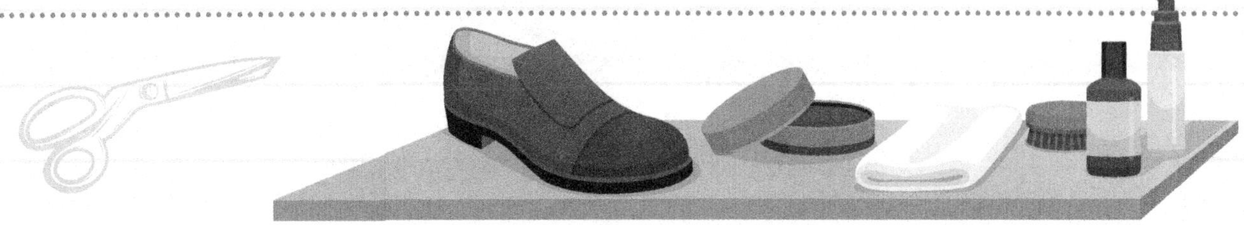

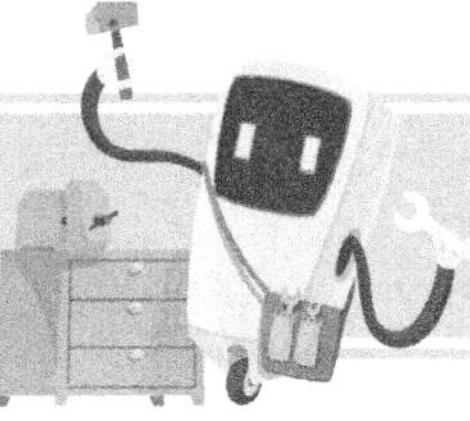

Read the poem and answer the questions that follow.

The Railway Train

By Emily Dickinson

I like to see it lap the miles,
And lick the valleys up,
And stop to feed itself at tanks;
And then, prodigious, step
Around a pile of mountains,
And, supercilious, peer
In shanties by the sides of roads;
And then a quarry pare
To fit its sides, and crawl between,
Complaining all the while
In horrid, hooting stanza;
Then chase itself down hill
And neigh like Boanerges;
Then, punctual as a star,
Stop — docile and omnipotent —
At its own stable door.

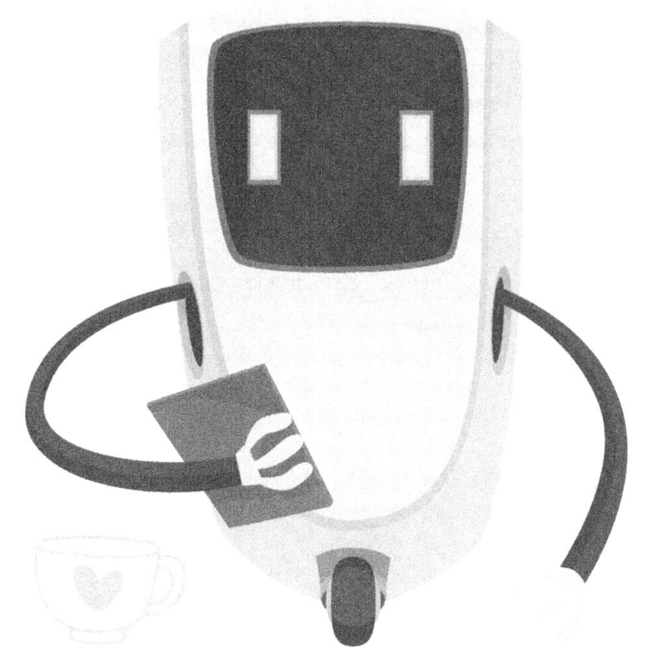

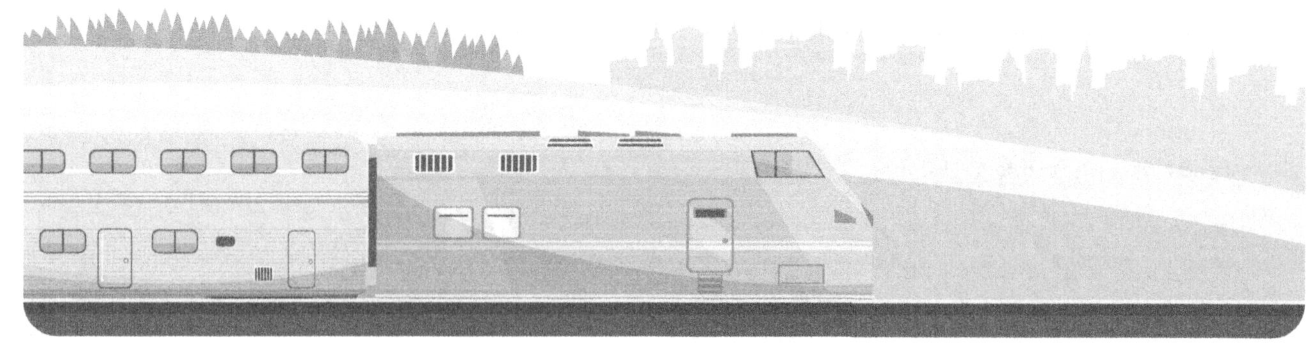

6. What does Dickinson describe over the course of the poem?

7. Explain what the train is being compared to.

8. What do you notice about the syntax or structure of the poem?

9. Why do you think Dickinson wrote this poem?

Let's get some fitness in! Go to page 237 to try some fitness activities.

FITNESS TIME

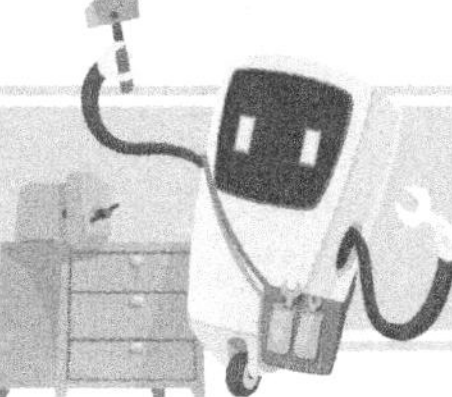

Part 1: Rhetorical Appeals

What Are Rhetorical Appeals?

Rhetorical appeals or rhetorical strategies describe the appeal or push to the audience's emotions or thoughts. Rhetorical appeals are in commercial ads, magazines, and social media. You might also find them on the internet. Rhetorical appeals can also be found in writing as well. There are three rhetorical appeals.

Key Terms

Ethos - a persuasive appeal using a person's character

Explanation - Ethos pinpoints whether or not a speaker or a person is trustworthy or has authority. Ethos shows or persuades the audience that they can trust a person or company. It depends on trusting a person or a specific brand who is selling a product. If there is a mention of a trusted person or organization, people will be more apt to listen or buy the product.

Example - an example of ethos in a commercial is when directors choose dentists to sell toothpaste. People are more apt to listen to a dentist when buying toothpaste than just a regular actor.

Pathos - a persuasive appeal to a person's emotions

Explanation - Pathos focuses on a person's emotions. When an ad pulls on the emotional heart strings, people will listen because they are emotionally invested.

Example - an example of a public service announcement that uses pathos would be an ASPCA commercial. The commercial might show footage of hungry and abused dogs while playing a sad song in the background. This pulls at an audience's emotions and coaxes them into wanting to help the cause.

Logos - a persuasive appeal to a person's logic or intelligence

Explanation - Logos focuses on intelligence and reasoning. It presents an audience with evidence and specific facts. People are provided charts, graphs, facts, statistics, and studies. An audience will be more apt to listen if there are researched facts about a claim.

Example - when selling a prescription drug, the makers of the commercial will include statistics about how much the pill has helped individuals.

Read the following sentences. Identify the appeal and explain why it is that particular appeal.

1. Colgate toothpaste is a trusted brand of toothpaste by dentists..

....

2. The study of Entresto shows it keeps patients' hearts healthy and keeps these patients out of the hospital.

3. There are shivering puppies that haven't eaten in weeks with their ribs sticking out.

4. Tara from Texas drank these shakes on the Atkins diet, and she lost fifty pounds in six months.

5. Studies show that fast food increases the risk of heart disease and obesity.

6. Do you want adventure? Do you want to snorkel, go on roller coasters, and zip line? Then, Disney World is for you!

7. The FDA approves this drug to help your heart become healthier.

8. Recent research studies show that since cell phone use among teenagers has increased, so have instances of educational plagiarism.

9. Make sure you lock your doors. There have been many break-ins, and two people were kidnapped last week.

10. Write a couple of sentences that use logos. Use additional paper to answer questions 10-12.

11. Write a couple of sentences that use pathos.

12. Write a couple of sentences that use ethos.

Part 2: Rhetorical Fallacies

What is a logical fallacy?

A logical fallacy is a weak claim. It gives incorrect or irrelevant information that does not fully explain the claim. However, sometimes logical fallacies are written so well they are not always easily seen as being wrong. Logical fallacies take away from a person's credibility when writing.

Key Terms

Ad Hominem - this is a personal attack in order to discredit someone

- Example - You would never understand my situation since you never had to struggle like I have.

  Write your own example - ....

  ....

Appeal to Nature - argument that focuses on the fact that natural choice is best

- Example - You should never take antibiotics. They are not good for you because they are unnatural.

  Write your own example - ....

  ....

Bandwagon Appeal - argument that shows that the popular choice is the best choice

- Example - Everyone has the latest iPhone. Why don't you?

  Write your own example - ....

  ....

Circular Reasoning - supports the claim by repeating the claim; goes in circles

- Example - I am practically a grown up. You should let me have a later curfew. You should let me come home at midnight because I am a grown up.

Write your own example - ....

....

Guilt by Association - unfair negative relationship with someone that makes you seem bad

- Example - I see that your grandfather smokes. I shouldn't be around you. I bet you smoke, too.

Write your own example - ....

....

Red Herring - gives an unrelated topic as a distraction

- Example - "Junior, why did you exceed your phone data this month?" asked Junior's mother. "Mom, you look very pretty today," exclaimed Junior.

Write your own example - ....

....

Hasty Generalization - a broad claim based on assumptions or stereotypes

- Example - Sara and Bobby failed the math test today. Looks like I will fail the test, too.

Write your own example - ....

....

Let's get some fitness in! Go to page 237 to try some fitness activities.

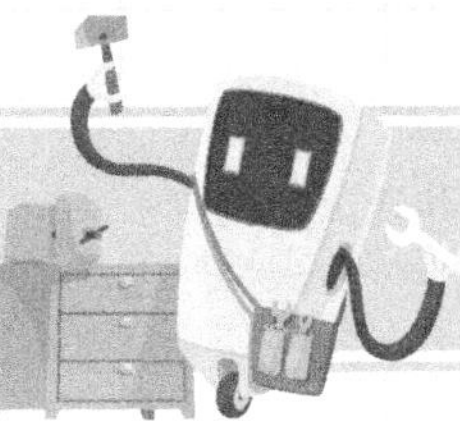

Using the Formula for Slope

Review:

Given two coordinate points on a line, we can determine the slope of the line. Considering our two coordinate points are (x_1, y_1) and (x_2, y_2). The formula we use to find the slope is:

$$\text{Slope} = (y_2 - y_1) / (x_2 - x_1)$$

Question 1:

What is the slope for a line crossing through the following coordinates: (0, 2) and (1, 4)?

A. $\frac{2}{1}$

B. $\frac{1}{2}$

C. $\frac{4}{1}$

D. $\frac{4}{0}$

Question 2:

What is the slope for a line crossing through the following coordinates: (1, 5) and (2, 10)?

A. $\frac{5}{10}$

B. $\frac{1}{2}$

C. $\frac{5}{1}$

D. $\frac{1}{5}$

Question 3:

What is the slope for a line crossing through the following coordinates: (0, 4) and (2, 0)?

A. 0

B. -4

C. $\frac{2}{1}$

D. $-\frac{2}{1}$

Question 4:

What is the slope for a line crossing through the following coordinates: (2, 4) and (4, 8)?

A. $\frac{4}{1}$

B. $\frac{2}{1}$

C. $\frac{1}{2}$

D. 4

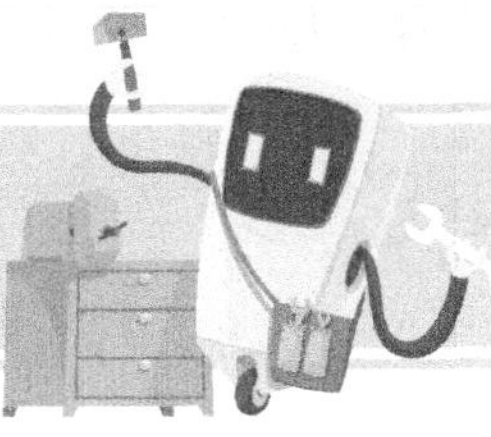

Question 5:

Given two coordinates on a line, Firestorm Warrior can find the slope of the line in the time it takes to scorch a marshmallow. For a line crossing through the coordinate points of (6,7) and (7,8), what did he find as the slope of this line?

Question 6:

After he tracked down graham crackers and chocolate, Firestorm Warrior found the slope of a line crossing through the coordinate points of (5,10) and (10,20). What is the slope of this line in its simplest ratio?

Question 7:

While making s'mores over a small flame in his open palm, Firestorm Warrior finds the slope of a line crossing through the coordinate points of (0,7) and (3,12). What is the slope of this line?

Question 8:

Sticky fingers don't keep Firestorm Warrior from figuring out the slope of a line crossing through the coordinate points of (2,20) and (3,30). What is the slope of this line?

Let's get some fitness in! Go to page 237 to try some fitness activities.

Comparing the Slopes of Linear Equations

Question 1:

Compare the slope of these two linear equations: $y = 2x$ and $y = 3x$. Which has a steeper slope?

A. $y = 2x$

B. $y = 3x$

Question 3:

Compare the slope of these two linear equations: $y = \frac{3}{2}x$ and $y = \frac{2}{3}x$. Which has a steeper slope?

A. $y = \frac{3}{2}x$

B. $y = \frac{2}{3}x$

Question 2:

Compare the slope of these two linear equations: $y = \frac{1}{2}x$ and $y = \frac{2}{3}x$. Which has a steeper slope?

A. $y = \frac{1}{2}x$

B. $y = \frac{2}{3}x$

Question 4:

Compare the slope of these two linear equations: $y = -2x$ and $y = -3x$. Which has a steeper slope?

A. $y = -2x$

B. $y = -3x$

Question 5:

While Firestorm Warrior roasts hot dogs over a flame shooting out of his big toe, he compares the slope of these two linear equations: $y = \frac{1}{2}x$ and $y = 2x$. Graph these two lines below and compare their slopes. Which line has a steeper slope?

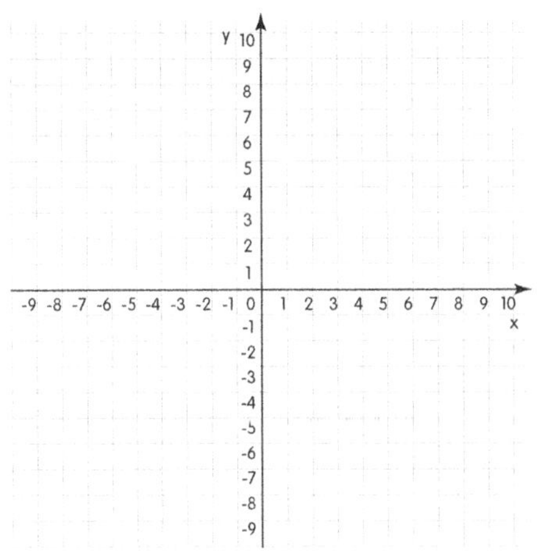

Question 6:

Firestorm Warrior hunts for ketchup and mustard while comparing the slope of these two linear equations: $y = -5x$ and $y = -10x$. Graph these two lines below and compare their slopes. Which one is steeper?

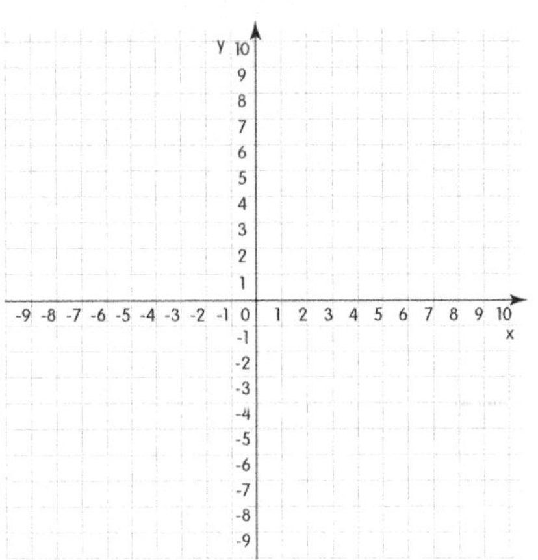

Question 7:

Firestorm Warrior is frustrated the number of hot dogs in one package does not match the number of hot dog buns in one package. To calm himself, he compares the slope of these two linear equations: $y = \frac{3}{2}x$ and $y = \frac{2}{3}x$. Graph these two lines below and compare their slopes. Which line has a steeper slope?

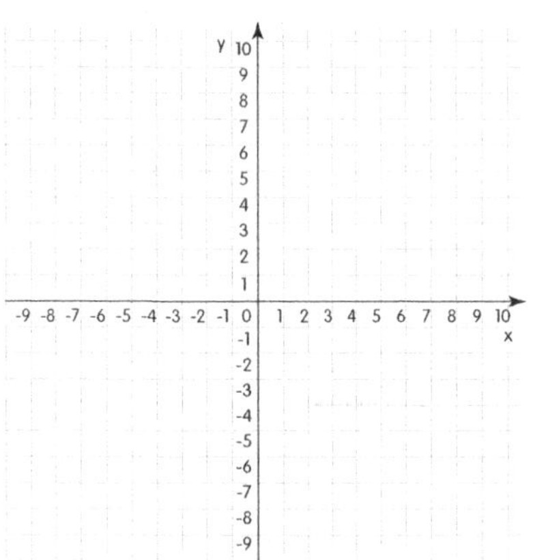

Question 8:

Firestorm Warrior loves sauerkraut and relish and heaps both onto his hot dog. Taking a bite, he compares the slope of these two linear equations: $y = \frac{2}{4}x$ and $y = \frac{3}{6}x$. Graph these two lines below and compare their slopes. Which line has a steeper slope?

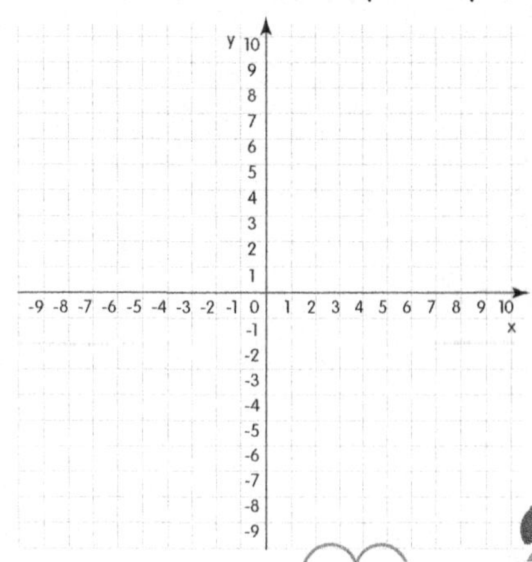

Let's get some fitness in! Go to page 237 to try some fitness activities.

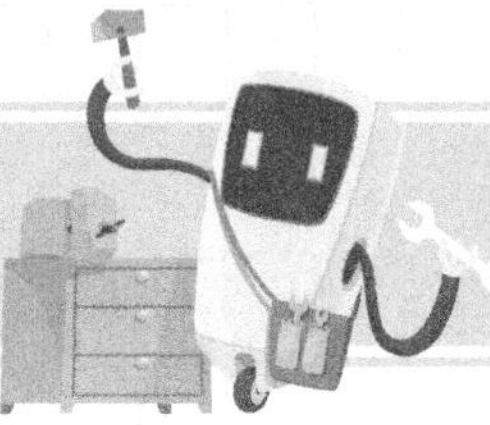

Solving Simple Equations with One Variable

Review:

To solve for a variable, add or subtract elements from both sides of the equation to isolate the variable on one side.

Question 1:

Solve for x in the equation: $5x + 4 = 7x$.

A. $x = 1$

B. $x = 2$

C. $x = \frac{1}{2}$

D. $x = 0$

Question 2:

Solve for b in the equation: $3b - 10 = 8b$.

A. $b = -2$

B. $b = 2$

C. $b = 3$

D. $b = 5$

Question 3:

Solve for m in the equation: $12m = 6m + 3$.

A. $m = \frac{1}{2}$

B. $m = 2$

C. $m = 1$

D. None of the above

Question 4:

Solve for y in the equation: $8y + 4 = 4y + 8$.

A. $y = 0$

B. $y = 4$

C. $y = 2$

D. $y = 1$

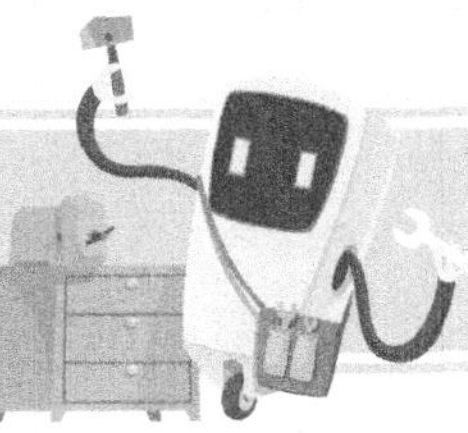

Question 5:

Firestorm Warrior loves a nice campfire. While sipping hot chocolate, he solves for h in the following equation: $10h + 5 = 25 + 5h$. What is the value of h?

Question 6:

Firestorm Warrior gazes up at the summer stars while solving for s in the following equation: $7s - 5 = 2s$. What is the value of s?

Question 7:

For a moment, Firestorm Warrior thinks he hears a bear in the woods. To stay calm, he solves for b in the following equation: $8b - 2 = 7b + 3$. What is the value of b?

Question 8:

Firestorm Warrior panics until he realizes the sounds were from an oversized chipmunk. He relaxes and solves for c in the following equation: $6c + 2 + 8 = 3c + 19$. What is the value of c?

Let's get some fitness in! Go to page 237 to try some fitness activities.

Fingerprints

SCIENCE EXPERIMENT

During gestation, the epidermis of the skin at the fingertips is formed. This means when a baby inside the womb is just twenty weeks old, they have their own unique fingerprint! Fingerprints contain ridges and patterns that are unique to a specific individual. Fingerprints never change as a person grows and ages. The size grows but not the pattern or the shape.

Fingerprints are used on a daily basis. Police locate fingerprints at a crime scene, people have their cellphones locked until an owner's fingerprint can unlock it, and some buildings require fingerprints for admittance. There are three different shapes to a fingerprint. The different shapes are a loop or curve, a whorl or circular pattern, and an arch.

Are fingerprint shapes genetic? Does DNA determine fingerprint patterns?

Let's complete an experiment to find out!

Materials Needed

Paper towel	White printer paper (several pieces)	Pencil
Clear tape	Scissors	A person you are related to
A person you are not related to		

Procedure

1. Rub a pencil on a piece of white paper. Make sure it is very thick. This will be your ink pad.
2. Then, rub your finger onto the pencil lead smudge, pressing down hard.
3. Now, take a clear piece of tape and stick it on your finger.
4. Place the tape on another white piece of paper, and label it with your name.
5. Ask a relative related by blood if you can take their fingerprint. Repeat the same process as you did with yours, and label it with their name.
6. Now, ask a friend or someone you are not related to by blood if you can take their fingerprint. Repeat the same process as you did with yours and your relative. Make sure you label it with their name.
7. After collecting all three samples, check to see if the pattern of your fingerprint matches in any way to the person you are related to.

Questions

1. Are the fingerprint patterns the same? Explain your results.

....

....

....

2. What does this experiment and its results tell you about genetics?

....

....

....

3. Explain how these results could help someone investigating a crime scene.

....

....

....

....

Let's get some fitness in! Go to page 237 to try some fitness activities.

Directions:
Go through the maze to help Harry find his birthday gifts.

WEEK 7

GRADE 8-9

In Week 7, you'll learn about rhyme schemes in a poem. We'll also go over fun words like onomatopoeia, anaphora, assonance, and alliteration.

Part 1 - Rhyme Scheme

A rhyme scheme points out the pattern or rhyme in a poem at the end of each line. To find the rhyme, assign each line a letter. If that particular line rhymes with another line, give it that same letter. Matching letters indicate a rhyme.

Here is an example:

Roses are red - A

Violets are blue - B

Sugar is sweet - C

So are you - B

In completing a rhyme scheme for this short poem, the first line that ends with the word "red" is labeled A. The second line ending with the word "blue" does not rhyme with line one, so it is given the letter B. The third line ending in "sweet" does not rhyme with either line one or line two, so it is given the letter C. The last line that ends with "you" rhymes with "blue" on line two, so it is given the same letter, which is the letter B. This short poem has an ABCB rhyme scheme.

Practice labeling the rhyme scheme with this next poem.

Clouds

By Anonymous

(Public Domain Material)

White sheep, white sheep, ...

On a blue hill, ...

When the wind stops, ...

You all stand still. ....

When the wind blows, ...

You walk away slow. ....

White sheep, white sheep, ...

Where do you go? ....

What is the rhyme scheme for this poem? ....

Part 2 - Poetry Terms

When analyzing poetry, it is important to be able to pick out poetry terms and see how they effect the overall mood of the poem. Understanding these poetry terms and how they relate to a poem will help you comprehend the poem on a much deeper level.

Key Terms and Practice

Anaphora is the repetition of words. These can be words at the end of the sentence or anywhere within a poem.

In this excerpt, identify anaphora and its purpose in the poem.

"Out of the cradle endlessly rocking, / Out of the mocking-bird's throat, the musical shuttle, / Out of the Ninth-month midlight." from *Out of the Cradle Endlessly Rocking* by Walt Whitman (Public Domain Material)

What is the anaphora in the excerpt? ....

What is its purpose? ....

Onomatopoeia are sound words. Sometimes, these words can be made up.

In this excerpt, identify onomatopoeia and its purpose in the poem.

"Tlot-tlot! Tlot-tlot! Had they heard it? The horse hoofs ringing clear." from *The Highwayman* by Alred Noyes (Public Domain Material)

What is the onomatopoeia in the excerpt? ....

What is its purpose? ....

Alliteration is the repetition of beginning consonant sounds. Remember that consonants are letters that are not vowels.

In this excerpt, identify the alliteration and its purpose in the poem.

"Where the wild men watched and waited / Wolves in the forest, and bears in the bush." from *A Night With a Wolf* by Bayard Taylor (Public Domain Material)

What is the alliteration in the excerpt? ....

What is its purpose? ....

Assonance is the repetition of vowel sounds.

In this excerpt, identify the assonance and its purpose in the poem.

"I wandered lonely as a Cloud / That floats on high o'er Vales and Hills / When all at once I saw a crowd, / A host of golden Daffodils." from *Wandered Lonely as a Cloud* by William Wordsworth.

What is the assonance in the excerpt? ....

What is its purpose? ....

Read the following poems. Identify the rhyme scheme and other poetry elements.

Theme in Yellow

By Carl Sandburg

(Public Domain Material)

I spot the hills ....
With yellow balls in autumn. ....
I light the prairie corn fields ....
Orange and tawny gold clusters ....
And I am called pumpkins. ....
On the last of October ....
When dusk is fallen ....
Children join hands ....
And circle round me ....
Singing ghost songs, Ooooo ....
And love to the harvest moon; ....
I am a jack-o'-lantern ....
With terrible teeth ....
And the children know ....
I am fooling ....

1. What is the anaphora used and what is its purpose? ....

....

2. What is the onomatopoeia used and what is its purpose? ....

....

3. What is the alliteration used and what is its purpose? ....

....

4. What is the assonance used and what is its purpose? ....

Who Has Seen the Wind?

By Christina Rossetti

(Public Domain Material)

Who has seen the wind? ....
Neither I nor you. ....
But when the leaves hang trembling, ...
The wind is passing through. ....
Who has seen the wind? ....
Neither you nor I. ....
But when the trees bow down their heads, ...
The wind is passing by. ....

5. What is the anaphora used and what is its purpose? ....

....

6. What is the alliteration used and what is its purpose? ....

Let's get some fitness in! Go to page 237 to try some fitness activities.

Part 1 - Active and Passive Voice

Active and passive voice refers to the tense of the verbs used in writing. In **active voice**, the object or the word that receives the action is written to show that the subject receives the action. The order used in active voice is the subject, the verb, and the direct object.

Here is an example: Jimmy ate the candy bar.

This sentence is written in active voice. Be careful! Some people think that active voice means that the verb always has to be in the present tense, that is not the case. In the example sentence above, the subject is "Jimmy," the verb is "ate," and the direct object is the "candy bar."

Passive voice, which is less common and should not be used in formal writing, changes up the order used in the example above. The common order used in passive voice is usually the direct object, the verb, and the subject.

Here is an example: The candy bar was eaten by Jimmy.

In passive voice, the subject receives the action. In the example above, the direct object is the "candy bar," the verb phrase is "was eaten," and the subject is "Jimmy."

The following sentences are written in active voice. Change them to passive voice so you can recognize the difference.

1. The Smiths redecorated their living room in order to attract people to purchase their house.

....

2. Millions of families visit Disney World every year. ....

....

3. Julie ate the last piece of cake. ....

....

The following sentences are written in passive voice. Change them to active voice so you can recognize the difference.

4. The pamphlet was read by the entire class. ....

....

5. The computer was used by John and it is now broken.

....

6. His homework was completed before the big game.

....

In the following sentences, identify the voice. If the sentence is written in passive voice, change it to active. If it is written in an active voice, explain why it is an example of active voice.

7. The new book was being read by the librarian. Active or Passive

....

8. The research at the conference was presented by Bill McGuire. Active or Passive

....

9. The old furniture was eventually picked up by the cleaning crew. Active or Passive

....

10. Angie wrote a letter to Santa before Christmas. Active or Passive

....

11. The winning touchdown was made by Randy, the running back. Active or Passive

....

12. Arnold shot the three pointer at the last second. Active or Passive

....

Part 2 - Parallel Structure

Parallel structure is not the repetition of words but instead the repetition of the structure of the sentence, or how the sentence is formed. Parallel structure uses the same pattern in order to make a point and grab the reader's attention. It also allows the text to have balance, since the structure remains equal.

Here is an example: Sarah ordered the flowers, picked up the decorations, and helped us set up for the party.

This sentence is an example of parallel structure because it visually shows the reader that Sarah will accomplish three things. She will order flowers, pick up decorations, and help set up. This structure is parallel due to the verb form of the word "will" and the commas between the tasks.

Determine if the following sentences have parallel structure. If it does contain parallel structure, then explain why. If the sentence does not contain parallel structure, then rewrite the sentence using parallel structure.

1. Amy enjoys swimming, surfing, and boogie boarding each summer on the ocean.

....

2. The secretary had to write letters quickly, accurately, and with a lot of detail.

....

3. Elvira was excited to go to the party and eat pizza. She was also excited to dance.

....

4. The researchers at the lab studied viruses. They studied cancer. They also studied blood borne pathogens.

....

....

5. Annie likes basketball, baseball, and soccer.

....

6. The team wanted to be more accurate and compassionate.

....

7. Mrs. Angel is evil-spirited, demanding, and despondent.

....

8. The peaches growing on the tree are ripe, orange, and sweet.

....

9. My sister Monica wants to go to New York and study fashion design.

....

10. The tourists visited the Golden Gate Bridge and ChinaTown.

....

11. My brother wants to see the monkeys at the zoo and the donkeys.

....

12. Our principal wants us to be nice, polite, and respectful.

....

13. The pet bird ate his seed, perched on the branch, and played with his toys.

....

14. Parrots come in many different colors like green, blue, and red.

....

15. When baby coconut crabs are little, they ride on top of their mother's backs.

....

Let's get some fitness in! Go to page 237 to try some fitness activities.

Linear vs. Nonlinear Functions

Review:

A linear function is a rule in which one input (x) produces only one possible output (y). A non-linear function is a rule in which an input (x) will have more than one possible output (y).

Question 1:

$y = 3x$ is a linear function.

True or false?

Question 2:

$y = \sqrt{25}$ is a linear function.

True or false?

Question 3:

$y = 6x + 2$ is a linear function.

True or false?

Question 4:

$y = x$ is a linear function.

True or false?

Question 5:

Firestorm Warrior's sidekick, Fire Hydrant, loves a hot summer day. He also loves linear functions. Does the table below represent a linear function? Explain.

x	y
1	-1
1	-2
1	-3
1	-4

Question 7:

When the fire truck shows up, Fire Hydrant gets excited. He shows the firefighters the table below. Does it represent a linear function? Explain.

x	y
0	0
0	-1
-1	0
-1	-1

Question 6:

Fire Hydrant floods the street by accident. In the midst of the chaos, he takes a look at the table below. Does the table represent a linear function? Explain.

x	y
1	-1
2	-2
3	-3
4	-4

Question 8:

The firefighters frown at Fire Hydrant and attempt to shut him off. He tries to distract them with another table. Does it represent a linear function? Explain.

x	y
1	1
2	4
3	6
4	8

Let's get some fitness in! Go to page 237 to try some fitness activities.

Graphs of Functions

Question 1:

Which set of coordinates represents a linear function?

A. (0, 0), (0, 1), (0, 2), (0, 3)

B. (1, 2), (1, 3), (2, 3), (3, 4)

C. (1, 2), (2, 3), (3, 4), (4, 5)

D. (1, 2), (2, 3), (3, 4), (3, 5)

Question 2:

Which set of coordinates represents a linear function?

A. (0, 1), (1, 1), (2, 2), (3, 3)

B. (2, 2), (2, 3), (2, 4), (2, 5)

C. (1, 2), (2, 2), (2, 4), (4, 5)

D. (1, 2), (1, 3), (3, 4), (3, 5)

Question 3:

Which set of coordinates represents a linear function?

A. (5, -1), (6, -2), (7, -3), (8, -4)

B. (1, 2), (1, -2), (2, 3), (3, -4)

C. (2, 2), (3, 3), (2, -2), (3, -3)

D. (1, 2), (1, 3), (1, 4), (1, 5)

Question 4:

Which set of coordinates represents a linear function?

A. (0, 0), (0, 1), (0, 2), (0, 3)

B. (1, 2), (1, 3), (1, 4), (1, 5)

C. (2, 2), (2, 3), (2, 4), (2, 5)

D. (1, 2), (3, 4), (5, 6), (7, 8)

Question 5:

As the sun beats down, Fire Hydrant distracts himself by looking at the two lines on the graph below. Which one represents a non-linear function? Explain.

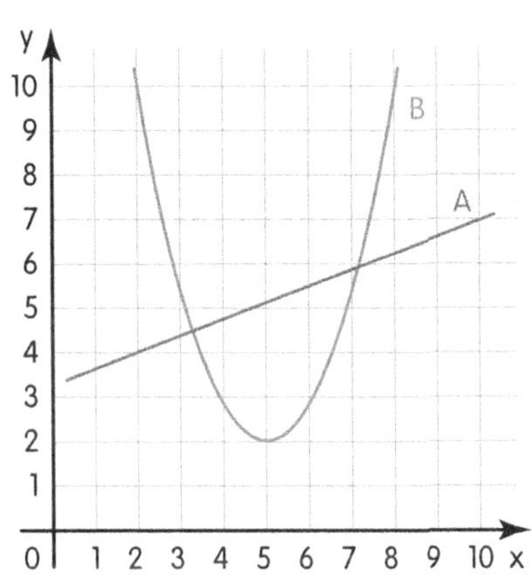

Question 6:

Fire Hydrant begins a steady drip of water from one valve. He studies another set of lines on a graph. Are they both linear or both nonlinear? Explain.

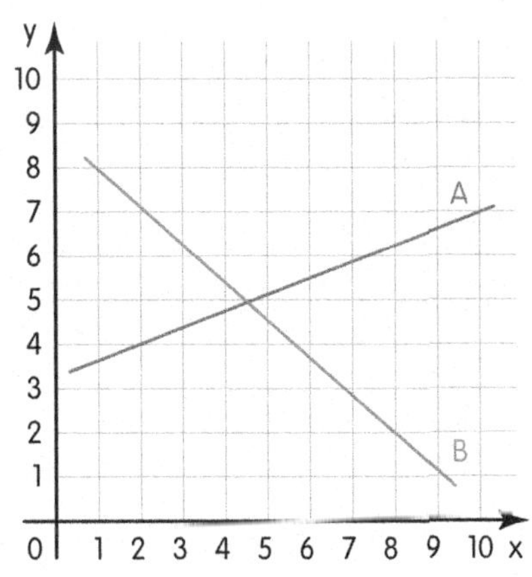

Question 7:

Fire Hydrant's steady drip has become a trickle. He studies yet another line on a graph. Is it linear or nonlinear? Explain.

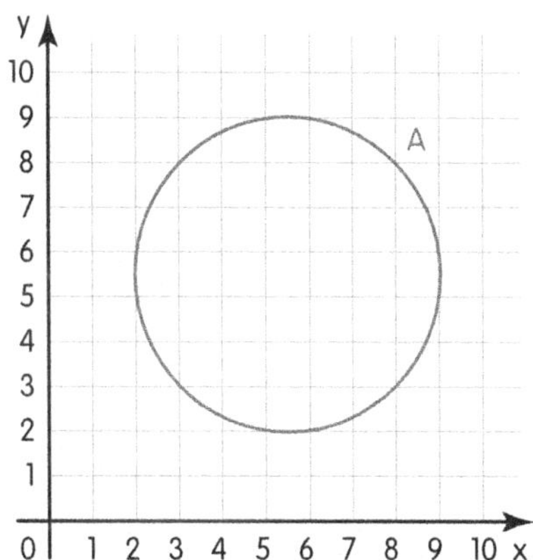

Question 8:

The trickle coming from Fire Hydrant is now a steady stream. When he hears a firetruck siren in the distance, he studies another line. Is it linear or nonlinear?

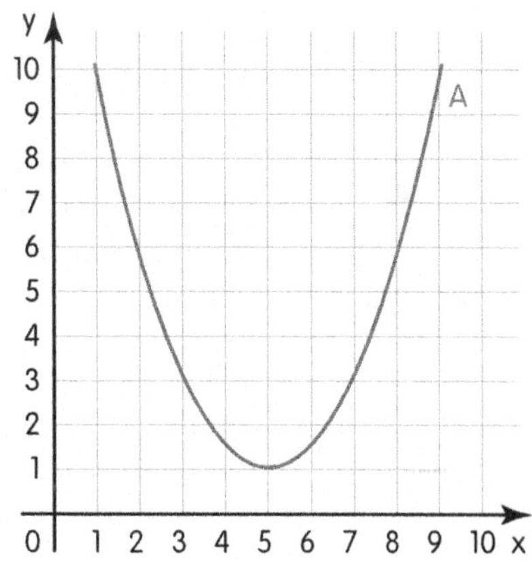

Let's get some fitness in! Go to page 237 to try some fitness activities.

Constructing Functions

Question 1:

Choose the linear function that reflects these coordinates: (1,1), (2,4), (3,9), (4,16).

A. $y = 2x$

B. $y = x + 2$

C. $y = 2x + 2$

D. $y = x^2$

Question 2:

Choose the linear function that reflects these coordinates: (1,6), (2,7), (3,8), (4,9).

A. $y = 6x$

B. $y = x + 5$

C. $y = 6x + 5$

D. $y = x - 5$

Question 3:

Choose the linear function that reflects these coordinates: (1,3), (2,5), (3,7), (4,9).

A. $y = 2x + 1$

B. $y = x + 2$

C. $y = 2x + 2$

D. $y = 2x$

Question 4:

Choose the linear function that reflects these coordinates: (1,1), (2,4), (3,7), (4,10).

A. $y = 3x - 2$

B. $y = 2(x + 3)$

C. $y = 2x + 2$

D. $y = 3x + 2$

Question 5:

After their third visit to shut him off, the fire department disciplines Fire Hydrant by leaving him with homework regarding linear functions. Secretly, he loves it. He works out the linear function for the table of values below. What is it?

x	y
1	6
2	11
3	16
4	21

Question 6:

Still working away happily, Fire Hydrant determines the linear function for the table of values below. What is it?

x	y
1	$1\frac{1}{2}$
2	2
3	$2\frac{1}{2}$
4	3

Question 7:

Stopping only to wipe the sweat off his steel, Fire Hydrant figures out the linear function for the table of values below. What is it?

x	y
1	1
2	8
3	27
4	64

Question 8:

Fire Hydrant finally works out the linear function for the table of values below and sprays a geyser of water in delight. What is it?

x	y
1	1
4	2
9	3
16	4

Let's get some fitness in! Go to page 237 to try some fitness activities.

You Are What You Eat?

Have you ever heard the saying, you are what you eat?

This saying usually aligns with an individual's diet, meaning that if you eat healthy foods, you are generally healthy, and vice versa; if you eat unhealthy foods, you are most likely unhealthy.

While this saying is not necessarily true, there is a lot of contemporary research regarding junk food and the human brain. For example, in a 2020 study conducted by Penn State, eating too much junk food during adolescence can affect not only your physical health and your body, but also your brain.

The study revealed that junk foods affect an adolescents' brain by altering the way they think, learn, and retain information, in both the short and long term.

Leading researcher, Carl Ellison, explains that junk food can also lead to "depression and anxiety" in this particular age group. Ellison, a department chair and author of many studies, explains that processed foods high in sugar and fat are the key factors leading to mood changes.

Through studying over three thousand adolescents, Ellison discovered that poor choices were made due to high levels of junk food consumption. These poor choices resulted in bad hygiene, poor dietary habits, and thoughts of laziness and depression. Adolescent and teenage brains are not fully developed and thus are able to change and mold. The prefrontal cortex is not fully developed and does not combat urges. For example, it is a lot harder for adolescents and teenagers to say no to pop, candy, and chips.

Last year, Ellison conducted a study on mice. He and his colleagues fed one hundred mice high fatty foods and the other one hundred mice healthy foods. The mice that ate the fatty foods gained weight, but what was interesting was they also performed worse on memory tests in the form of mazes. After working with mice, Ellison then tested humans to confirm the results.

1. What is the purpose of the prefrontal cortex?

....

2. What is the main idea of this article?

....

3. What should adolescents and teens do with this newfound information?

....

4. What should be done to increase information about the physical and mental harms of junk food?

....

5. What might be some questions to ask about Ellison's research with the mice?

....

6. Why do you think junk food is so popular?

....

7. According to the article, why is it harder for adolescents and teenagers to say no to junk food?

....

Let's get some fitness in! Go to page 237 to try some fitness activities.

Directions:
Where are my clothes?

Answer:

ABC

WEEK 8

GRADE 8-9

In Week 8, you will learn how to properly cite pieces of works in your writing. In the math section, you will work on problems dealing with rotations, reflections, and translations.

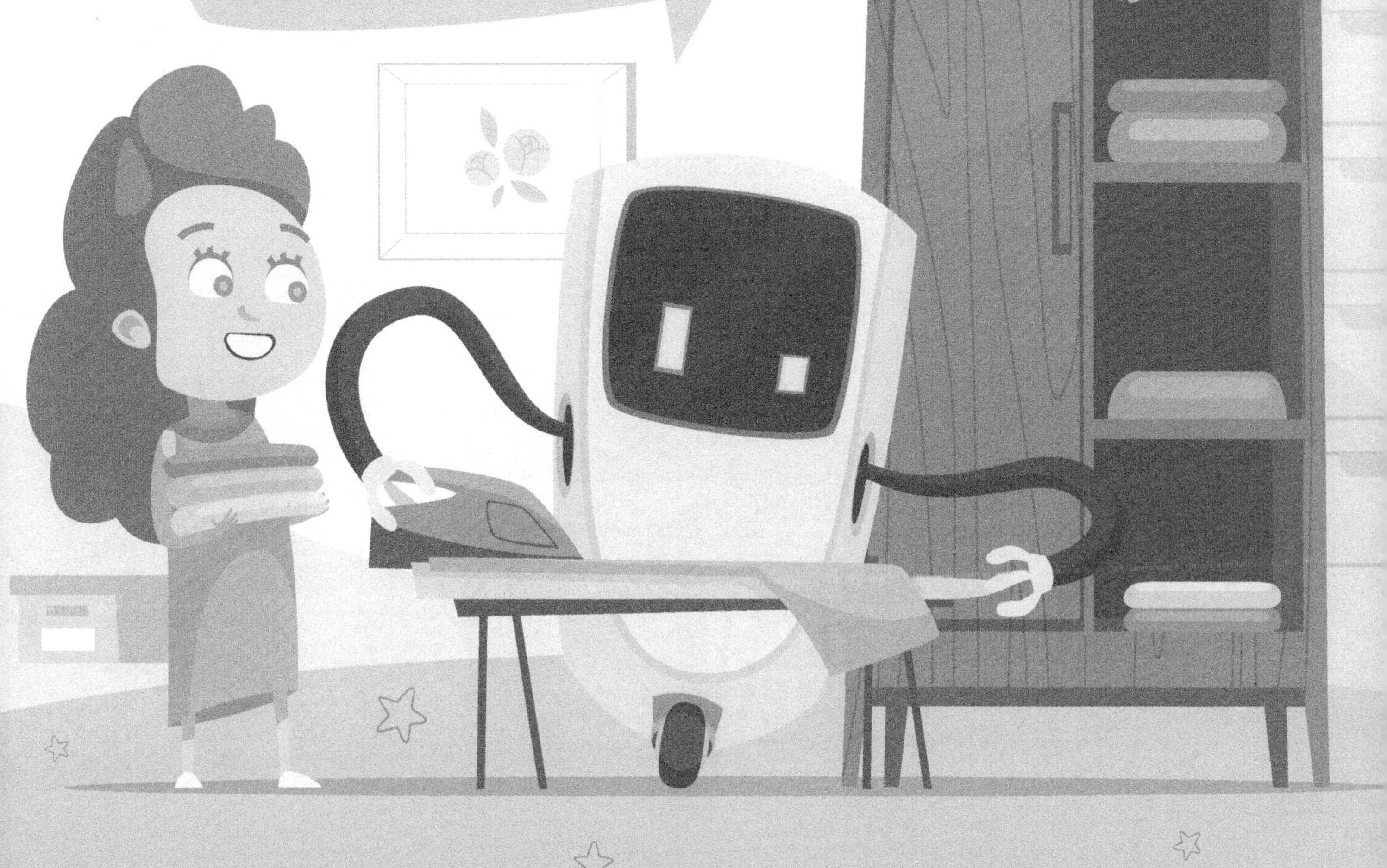

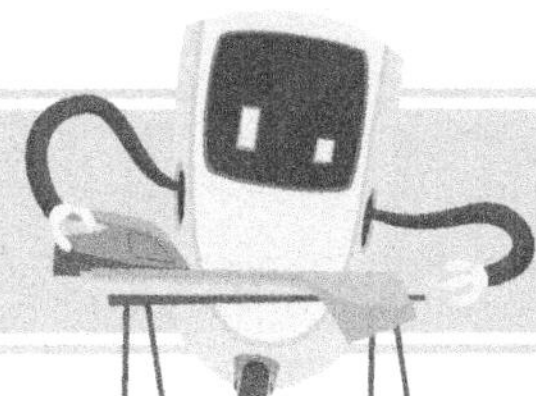

Key Terms

Main Claim - the author's main argument or what the author is trying or wanting to prove

Supporting Claims - sentences that support the author's main claim and give reasons why the main claim is correct

Evidence - researched and factual information that backs up the supporting claims and main claim

Counterclaims - claims that disagree with the main claim

In a persuasive or an argumentative text, it is important to understand all elements in order to understand the argument the author is making. The main claim is what the author wants to prove in the text, and they will use persuasion in order to make you think, feel, and believe they are correct in their thinking. They will hold up their main claim with supporting claims or details which explain and support what they are trying to prove. Authors of persuasive and argumentative texts will also include evidence showing they have researched and are credible regarding the particular topic they are writing about. Without evidence, there will not be enough information to prove their side. They need to convince the readers they are correct, and evidence can help do this. Authors often bring up a counterclaim in order to show they fully thought about what they are explaining and have acknowledged the other side. They also bring up the counterclaim in order to refute it and prove it is weak.

Read the following passage. Then, identify the persuasive and argumentative elements in the text.

Meditation

Schools are continuing to improve in many areas. Many schools have their students practice writing in every classroom, including math and science, in order to improve their written communication skills. Schools also incorporate more hands-on activities, and many are adding wellness topics into everyday subjects like physical education, history, and science. Many schools have started to implement meditation into their classes because they find many benefits of it for students of all ages. An example of including meditation into classes is having elementary students take a moment of silence to relax their bodies after recess. Many educators think it is important for students to take the time to think about what they are feeling. Experts feel that meditation can help students of all ages reduce stress, which in turn will help students reach more academic milestones. Schools in the United States should teach their students how to meditate.

There are many schools across the United States and Europe that are already implementing meditation into the school day and the results have been positive. In one study conducted by Indiana University, researchers discovered that elementary and middle school students were more focused after an eight-week term of meditation inserted into the class day for fifteen minutes. They also witnessed reluctant students becoming more inclined to participate. The quiet students gained confidence from the meditation lessons and began talking with their peers and participating in cooperative learning. At Woodard Elementary School in Indianapolis, the attendance rate skyrocketed after eight weeks of meditation classes. Their Dibels and Math XL tests also started to improve. At the beginning of the eight weeks, the fourth grade students were at a 63% proficiency level in reading and a 49% proficiency level in math. However, after sixteen weeks of implementing meditation classes, the scores grew to 89% and 73% respectively. This study shows that with meditation programs, the students at Woodard Elementary are reaping the benefits of increased attendance and higher academic scores. Additionally, after Leland Middle School in Lafayette, Indiana implemented meditation into their curriculum for the entire school year, they increased their scores on the ISTEP standardized test by eleven percent in English and seventeen percent in math.

Meditation programs should be included in every school across the United States, because they help students live healthier lives. The University of California, Los Angeles, conducted another study to see if meditation classes really helped students. After implementing a survey at Titus Elementary school in Los Angeles, the authors of the study wanted to understand how parents, administrators, teachers, and students were feeling and thinking after six months of meditation classes. The participants of the survey were in favor of meditation and believed in its benefits. Administrators, teachers, parents, and students indicated they noticed that meditation classes helped them reduce and manage stress, skills that could be used for the rest of their lives. It was also indicated that students diagnosed

with anxiety were able to control it better through the help of meditation classes. The survey results revealed that eighty-nine percent of the students felt less stressed in and out of school, and that the meditation program helped them feel this way. After adding in a yoga class at Parks High School in Sacramento, California, the high school students, both males and females, felt more relaxed and less stressed. The leading researcher, Alvin Honeycutt, explained, "This meditation program not only makes the students less stressed, but it gives them the tools to reduce stress in their own lives. If they are experiencing stress at home, they can use something they learned in meditation class to help them. They can also extend this knowledge to people in their family as well" (Honeycutt 45).

Even though the meditation program is positive and should be implemented in every school across the United States, some people are not convinced. Many individuals feel there is not enough long term research on the subject to genuinely know if there is a benefit to this program. Bart Bomgarder, Head Sociology Professor at Duke University, explains that "Research on this program is in its early stages," but he also says that "there is positivity seen in children and this program should continue" (Bomgarder 198). While Bomgarder feels the research is in its early stages, he does see the positives of this program. However, other individuals feel that meditation is not a good use of time and that students should instead be working on equations and reading. They think meditation class wastes time; however, they are wrong. Meditation helps improve academic performance in students and should be continued.

Even though incorporating meditation into schools is a new idea, there are many benefits that are already visible for students elementary to high school age. These students have learned how to cope with stress and how to ease and lessen their anxiety. Incorporating meditation classes into schools allows students to concenrate in school. They also improve attendance and help student grow academically. Meditation classes should be part of every students' life at every school in the United States.

1. What is the author's main claim?

....

....

2. What is one supporting detail that supports this claim?

....

....

3. What is one piece of evidence that supports this claim?

....

....

4. What is another supporting detail that supports this claim?

....

....

5. What is one piece of evidence that supports this supporting claim?

....

....

6. What is one counterclaim given? Was it strong? Explain.

....

....

7. What is another counterclaim given? Was it strong? Explain.

....

....

Let's get some fitness in! Go to page 237 to try some fitness activities.

FITNESS TIME

Part I - Works Cited Page

When you write a text with evidence, your teacher will require you to have a Works Cited Page. This particular page is inserted after the essay and contains information about where you found your evidence. The Works Cited Page has to be arranged based on a set of guidelines from the Modern Language Association, also known as MLA format. When citing your information on a Works Cited Page, you need to follow this particular order:

Author. Title of Source. Title of the Larger Source or Piece. Version or Number (if applicable). Publisher, Publication Date. Date of Access (if found on the Internet).

Some notes about this format:

- The author's name is listed with the last name first, a comma, and then the author's first name followed by a period. If there is more than one author, then you will list them all, adding a comma in between the names.
- The title of the source is put in quotations marks, followed by a period.
- The title of the large source is the magazine or website title. This is put in italics and followed by a period.
- If applicable and the source is a magazine or a journal, there will be a version number you will need to include.
- If your source is a book or something in print, you will include the publisher. If not, you will skip this step and include the date it was published.
- If your source was retrieved online, you will also include the date you retrieved it from the internet, followed by a period.
- If you cannot locate a certain piece of information like the author's name (this would be absent if the source was written by an organization) or a date, then you just skip that step and move on to the next needed element.

Also note - when the source takes up the second line, you have to include a hanging indent. This means the second line is tabbed over, which indicates that it belongs to the line above.

Here is an example of a Works Cited Page entry from a famous novel:

Austen, Jane. *Pride and Prejudice*. Dover Publications, 1995.

Please practice using the Works Cited Page form. Find three books around your house or school and practice making a Works Cited Page entry.

1. ....

2. ....

3. ....

Look at the following sources for published entries. What is missing or what is incorrect?

4. Louisa May Alcott. *Little Women*. Penguin Books, 1953.

....

5. Orwell, George. 1984. London: Secker and Warburg, 1949.

....

6. Austen, Jane. *Pride and Prejudice*. 1995. New York: Modern Library.

....

Now practice writing a Works Cited Page entry from an Internet Source. Find some articles about online education. Then create a Works Cited Page entry for each article that you found.

7. ....

....

8. ....

....

9. ....

....

Look at the following sources. What is missing?

10. "Outbreak of Lung Injury Associated with the Use of E-Cigarette, or Vaping, Products." *Smoking & Tobacco Use.*

11. Zubrzycki, Jaclyn. Year-Round Schooling Explained. *Education Week.* 18 December 2015. 23 January 2020.

12. "About Pets & People." Centers for Disease Control and Prevention. 15 April 2019. 23 January 2020.

Part 2: In-Text Citations

Not only do you need a Works Cited Page when using evidence in writing, but you also need to include in-text citations. Just like a Works Cited Page, in-text citations also require strict MLA formatting. You can cite evidence in one of two ways. The first is a **direct quote**. A direct quote includes word-for-word information from the source. This is usually indicated by a signal phrase or by information about the quote before the quotation. Then, you include the information in quotations, followed by a citation in parenthesis that includes the author's last name and the page number where the quote was found. An **indirect quote** does not use exact word for word information but instead includes the information in your own words. This format does still require a citation at the end to indicate where you found your paraphrased information.

Here is an example of a direct quote from a book - In Lois Lowry's *The Giver*, the author describes the apple's color as "nondescript" (Lowry 24).

Here is an example of an indirect quote from a book - In *The Giver*, Lois Lowry explains that the apple is not descriptive in color (Lowry 24).

1. Find a book in your house and write the Works Cited Page entry to it below.

....

Now include two indirect in-text citations from that book.

2. ....

3. ....

Include two direct in-text citations from that particular book.

4. ....

5. ....

Look at the citations below and explain what is wrong with them.

6. In thc election, the popular vote initially favored President Richard Nixon over his opponent George McGovern (Mound, 9).

....

7. In the book *Wonder*, Auggie is the main character (*Wonder*).

....

8. In the article entitled *The New Wave*, George Thomas explains that students are "stressed out due to having to complete school online, rather than seeing their friends and working together in person." (Thomas 45).

....

9. In *Four Ways To Stay Healthy Around Your Pet*, the author explains that it is important to not be around your pet if you have a cough (3.)

....

Let's get some fitness in! Go to page 237 to try some fitness activities.

Rotations, Reflections, and Translations

Review:

Congruent figures are identical in size and shape and can represent a rotation, reflection, or translation of each other. A rotation is a pivot of a shape in a circular motion, a reflection is a mirror image of a shape, and a translation is a move up/down/left/right of a shape.

Question 1:

What transformation do these two congruent figures represent?

A. A rotation
B. A reflection
C. A translation
D. None of the above

Question 2:

What transformation do these two congruent figures represent?

A. A rotation
B. A reflection
C. A translation
D. None of the above

Question 3:

What transformation do these two congruent figures represent?

A. A rotation
B. A reflection
C. A translation
D. None of the above

Question 4:

What transformation do these two congruent figures represent?

A. A rotation
B. A reflection
C. A translation
D. A and C

Question 5:

Adrastos the Super Warrior has arranged for two large billboards with his face on them. The first is 20 meters high and 50 meters wide. The second is 50 meters high and 20 meters wide. How are these two billboards congruent? Explain in terms of rotation, reflection, or translation.

Question 6:

Adrastos the Super Warrior has purchased two identical banners in the shape of an arrow. In his bedroom, he plans to display them side-by-side with one pointing to his portrait on the right and one pointing to his portrait on the left. How are these two arrows congruent? Explain in terms of rotation, reflection, or translation.

Question 7:

To celebrate his birthday, Adrastos the Super Warrior buys two enormous paper hearts identical in size. He sends them to himself. When he displays them on his refrigerator, one is above and one is below. How are these two hearts congruent? Explain in terms of rotation, reflection, or translation.

Question 8:

In the middle of summer, Adrastos the Super Warrior decides to set off fireworks in celebration of his own accomplishments. His favorite are the two identical fireworks in the shape of his profile—one looking left and one looking right. How are these two profiles congruent? Explain in terms of rotation, reflection, or translation.

Let's get some fitness in! Go to page 237 to try some fitness activities.

Comparing Similar and Congruent Figures

Review:

Congruent figures are identical in size and shape and can represent a rotation, reflection, or translation of one another. Similar figures are identical in shape but may be different sizes and have a different orientation. Similar figures will have identical corresponding angles and proportional corresponding side lengths.

Question 1:

How would you compare these two figures?

A. Congruent

B. Similar

C. Neither congruent nor similar

Question 2:

How would you compare these two figures?

A. Congruent

B. Similar

C. Neither congruent nor similar

Question 3:

How would you compare these two figures?

A. Congruent

B. Similar

C. Neither congruent nor similar

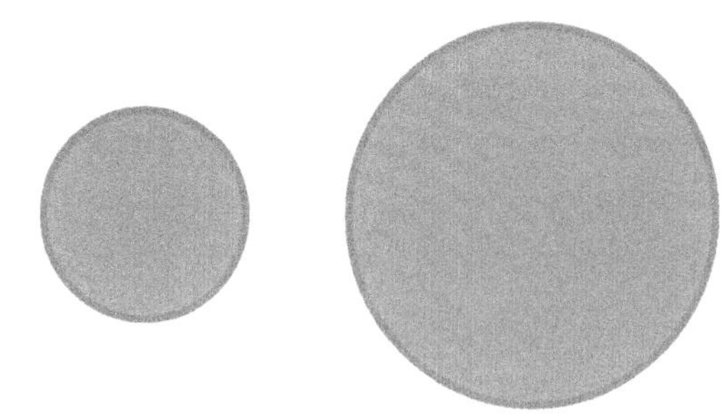

Question 4:

How would you compare these two figures?

A. Congruent

B. Similar

C. Neither congruent nor similar

Question 5:

To watch video clips of himself, Adrastos the Super Warrior buys a rectangular television screen for his apartment in the city. It measures 5 meters wide by 3 meters tall. He buys a smaller rectangular screen for his house on the beach. It measures 2.5 meters wide by 1.5 meters tall. Are these screens similar figures? Explain.

Question 6:

Adrastos the Super Warrior decides to buy a third television screen for his bathroom. It measures $1\frac{1}{2}$ meters wide by $\frac{1}{2}$ meter tall. Is this screen a similar figure to the screens in question 5? Explain.

Question 7:

After a long day of building a sand sculpture of himself, Adrastos the Super Warrior orders two 12-inch cheese pizzas. He eats half of one and falls asleep. At this point, are the two pizzas similar figures? Explain.

Question 8:

When it's time to rescue a whale stranded on the beach, Adrastos the Super Warrior orders two perfectly circular spotlights to focus on him. One has a diameter of 10 meters and the other has a diameter of 15 meters. Are these two spotlights similar figures? Explain.

Let's get some fitness in! Go to page 237 to try some fitness activities.

Similar Figures

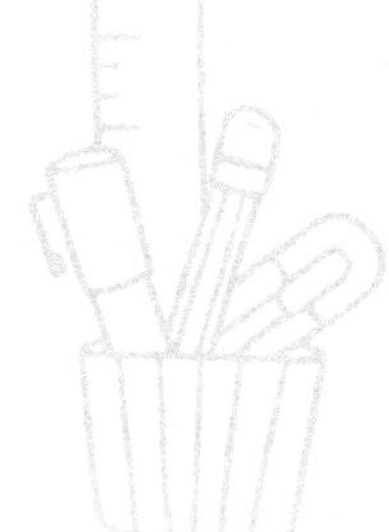

Review:

Similar triangles will have identical corresponding angles and proportional corresponding side lengths.

Question 1:

For these two triangles to be similar, what is the length of the missing side?

A. 8

B. 10

C. 6

D. 12

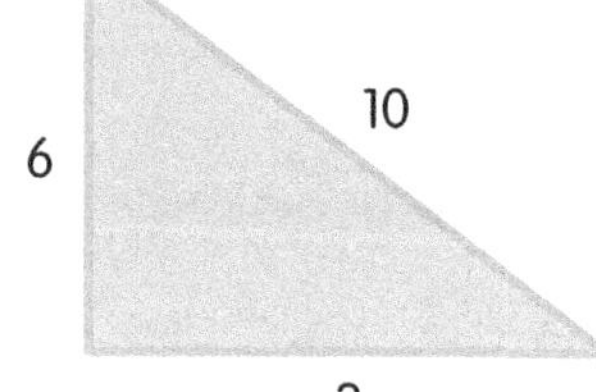

Question 2:

For these two rectangles to be similar, what is the length of the missing width?

A. 10

B. 20

C. 15

D. None of the above

Question 3:

For these two squares to be similar, what is a possibility for the length of the missing width?

A. 3

B. 4

C. 6

D. All of the above

Question 4:

For these two circles to be similar, what is a possibility for the length of the missing diameter?

A. 20

B. 30

C. 40

D. All of the above

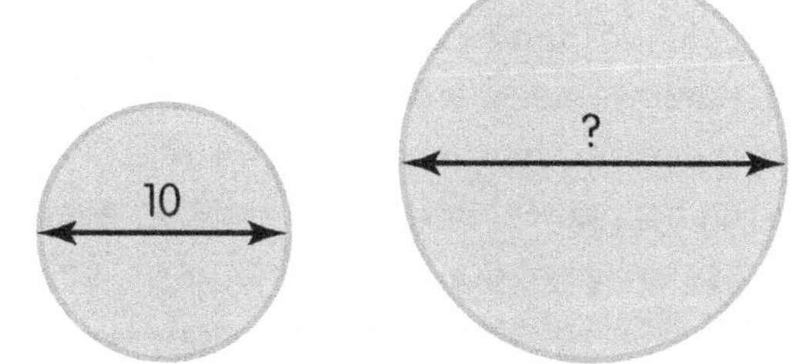

Question 5:

Adrastos the Super Warrior wants to have two similar rectangular buildings to feature a mural of him. The mural on the smaller building will be 60 meters wide by 120 meters tall. The mural on the larger building will be 90 meters wide. How tall will it be if this mural is a similar figure?

Question 6:

Adrastos the Super Warrior has chosen a third rectangular building for another mural. This mural will be a similar figure to the other two. If it's 240 meters tall, how wide is it?

Question 7:

For the pleasure of his fans, Adrastos the Super Warrior has ordered rectangular bumper stickers that are similar figures. The measurement of the first bumper sticker is 10 centimeters tall by 30 centimeters wide. The other bumper sticker measures 5 centimeters tall. What is the width?

Question 8:

Adrastos the Super Warrior passes out equilateral triangular flags with his logo down the beach. They come in two sizes. Would these flags be considered similar figures? Explain.

Let's get some fitness in! Go to page 237 to try some fitness activities.

What Does Every Living Organism Need?

Every living organism needs the basics, which are most commonly food and water. Animals and humans need water and the nutrients in food to produce energy. Plants do as well, but they receive their nutrients in a different way through the process of photosynthesis. The sun gives plants and flowers the energy they need to grow, develop, and reproduce. The process of photosynthesis requires several steps that provides a plant's cells the ability to take in the light and energy from the sun and change it into chemical energy. The plants use the light from the sun, water from the ground, and carbon dioxide in the air to produce sugar molecules the plants will use.

Now that you know about the process of photosynthesis, let's complete an experiment that will visually show you how this process works and how it keeps plants alive.

Materials Needed

2 transparent cups	Water	Baking soda
Liquid dish soap	Spinach leaves	Hole puncher

Procedure

1. Fill both cups with water.
2. Add a drop of dish soap to each cup and stir to dissolve. (This will keep the spinach leaves from wilting.)
3. Using the hole puncher, punch out thirty circles from the spinach leaves and place fifteen circles in each cup of water.
4. In the first cup, put in a half a cup of baking soda.
5. Place both cups in a window with sunlight.
6. After a day, go back to the two cups and view the results of the spinach leaves in the soapy water compared to the leaves in the baking soda water.

Questions and Reflection

1. Visually, how were the cups different? What did the spinach leaves in the cup of soapy water look like in comparison to the spinach leaves in the cup of baking soda?

....

....

2. How did the baking soda affect the rate of photosynthesis?

....

....

3. Why do you think the baking soda sped up the process of photosynthesis?

....

....

4. Why did the spinach leaves rise to the top of the cup with baking soda?

....

....

5. What did this experiment teach you about photosynthesis?

....

....

Let's get some fitness in! Go to page 237 to try some fitness activities.

MAZE

Directions: Using the graphic sequence below, go through the maze to the end point marked by the gear.

WEEK 9

GRADE 8-9

Week 9 takes a closer look into rhetorical strategies, prefixes, suffixes, roots, and pythagorean theorem. In the science section, you will learn about the discovery of water on the moon.

Have you ever heard of rhetorical strategies?

Rhetorical strategies are ways that writers use and manipulate language, like words and sentences, in order to achieve their purpose. This purpose can be to persuade, inform, or even to entertain. Authors especially use rhetorical strategies in historical fiction where they take a real event in history and spice it up with a fictional spin through diction, syntax, tone, and figurative language.

Read the following poem that was written in response to a real historical event. Then, answer the questions that follow.

Paul Revere's Ride

(Abbreviated Version)
by Henry Wadsworth Longfellow
(Public Domain Material)

Listen, my children, and you shall hear
Of the midnight ride of Paul Revere,
On the eighteenth of April, in Seventy-five:
Hardly a man is now alive
Who remembers that famous day and year.

He said to his friend, "If the British march
By land or sea from the town to-night,
Hang a lantern aloft in the belfry arch
Of the North Church tower as a signal-light,
One, if by land, and two, if by sea;
And I on the opposite shore will be,
Ready to ride and spread the alarm
Through every Middlesex village and farm,
For the country folk to be up and to arm."

Then he said, Good-night! and with muffled oar
Silently rowed to the Charlestown shore,
Meanwhile, his friend, through alley and street
Wanders and watches with eager ears,
Till in the silence around him he hears
The muster of men at the barrack door,
The sound of arms, and the tramp of feet,
Marching down to their boats on the shore.

Then he climbed to the tower of the Old North Church
To the highest window in the wall,
Where he paused to listen and look down
On a shadowy something far away,
Where the river widens to meet the bay, —
On the opposite shore walked Paul Revere.

Now he patted his horse's side,
Now gazed at the landscape far and near,
But mostly he watched with eager search
The belfry-tower of the Old North Church,
And lo! as he looks, on the belfry's height
A glimmer, and then a gleam of light!

He springs to the saddle, the bridle he turns,
A second lamp in the belfry burns!
Struck out by a steed flying fearless and fleet:
That was all! And yet, through the gloom and the light,
The fate of a nation was riding that night;
He had left the village and mounted the steep.

It was twelve by the village clock
When he crossed the bridge into Medford town.
He heard the crowing of the cock,
And the barking of the farmer's dog,
And felt the damp of the river fog,
That rises after the sun goes down.
It was one by the village clock,
When he rode into Lexington.
It was two by the village clock,
When he came to the bridge in Concord town.

You know the rest. In the books you have read,
How the British Regulars fired and fled, —
How the farmers gave them ball for ball,
From behind each fence and farm-yard wall,
Chasing the red-coats down the lane,
Then crossing the fields to emerge again
Under the trees at the turn of the road,
And only pausing to fire and load.
So through the night rode Paul Revere;
And so through the night went his cry of alarm
To every Middlesex village and farm, —
A cry of defiance and not of fear,
A voice in the darkness, a knock at the door,
And a word that shall echo forevermore!

1. What rhetorical strategies based on diction, otherwise known as word choice, did the author use?

....

....

2. Why did the author use this type of diction?

....

....

3. What was the tone of the poem?

....

....

4. How do you know this type of tone was used? What diction words reveal this?

....

....

5. Why do you think the author, Henry Wadsworth Longfellow, chose to take this historical event and write a poem about it?

....

....

6. Find one piece of figurative language that Longfellow used in this poem. Identify it and give an example.

....

....

7. Why did Longfellow use this type of figurative language? What was its purpose?

....

....

8. Who is the speaker of the poem?

A. A British soldier wanting revenge

B. Paul Revere himself

C. A good friend of Paul Revere reliving the moment

D. Someone telling the story of Paul Revere's ride

9. What is the poem's theme?

A. Many people died in this particular war.

B. War should never be a solution to a problem.

C. Perseverance and bravery in the face of danger result in lives saved.

D. Friendship is the most important relationship in life.

10. Read the following lines from the poem:

"Hardly a man is now alive
Who remembers that famous day and year."

What does this particular line mean?

....

....

11. Read this following line from the poem:

"For the country folk to be up and to arm."

Longfellow chose to use the word "arm." What does this mean and what particular rhetorical strategy did he appeal to?

....

....

12. Read the following lines from the poem:

"Wanders and watches with eager ears,
Till in the silence around him he hears
The muster of men at the barrack door."

What example of figurative language is this and what is its purpose?

....

....

13. Read the following lines from the poem:

"But mostly he watched with eager search
The belfry-tower of the Old North Church."

What rhetorical strategy is the author using here and what is its purpose?

....

Let's get some fitness in! Go to page 237 to try some fitness activities.

Part 1: Prefixes and Suffixes

What are prefixes and suffixes?

A **prefix** includes a letter or letters that are added to the beginning of a word in order to create a new word. A **suffix** is just the opposite. It is a letter or letters that are added to the end of a word in order to create a new word.

Why is it important to know different prefixes and suffixes?

It is important to know many different prefixes and suffixes because if you can recognize them, that knowledge can help you figure out a word that is unknown to you. The prefix or suffix can give you a clue about what the word means.

Prefixes

Here are some prefixes that will be helpful for you to know.

pre - before
re - again or back
sub - under, below, or less than
mis - incorrect or wrong
un - not or the opposite of
dis - not or the opposite of
in - not or the opposite of
im - not or the opposite of
non - not or the opposite of

1. Write two words that include the prefix pre - . ...

2. Write two words that include the prefix re - . ...

3. Write two words that include the prefix sub - . ...

4. Write two words that include the prefix mis - . ...

5. Write two words that include the prefix un - . ...

6. Write two words that include the prefix dis - . ...

7. Write two words that include the prefix in - . ...

8. Write two words that include the prefix im - . ...

9. Write two words that include the prefix non - . ...

Using the prefix information, try to figure out the definitions of the bolded words below.

10. The surgical cut the brain surgeon made was **precise**.

....

11. The man's hair was **receding** quickly after he reached the age of forty.

....

12. Lydia was a very **submissive** group member.

....

13. The gas station was **mismanaged** and, ultimately, closed down.

....

14. Jeremy was feeling **unwell** this morning, even though it was his birthday.

....

15. Aaron **disembarks** at three-thirty on Thursday.

....

16. Mario **intends** to do his homework between dinner and karate practice.

....

17. The villain in the video game is **immortal**.

....

18. The song my little sister learned in preschool is **nonsensical**.

....

Suffix

Here are some suffixes that will be helpful for you to know.

ful - having, showing, or causing
less - without
able - to be or do something
ible - to be or do something

19. Write two words that include the suffix - ful. ....

20. Write two words that include the suffix - less. ....

21. Write two words that include the suffix - able. ....

22. Write two words that include the suffix - ible. ....

Using the suffix information, try to figure out the definitions of the bold words below.

23. Lena was **boastful** after winning the state title. ....

24. Jillian thought that learning math was **useless**, so she never did her homework.

....

25. There is no help in sight for the **foreseeable** future. ....

26. My little brother is so **gullible**. ....

Part 2: Greek and Latin Roots

What is a Greek or Latin root?

These roots are from the Greek and Latin origin. A root is part of a word, mainly the base. Just like prefixes and suffixes, knowing a root word can help you understand what a word means.

Here are some roots that will be helpful for you to know.

bel - war
bene - good
fac - make or do

27. Write two words that include the root bel. ....

28. Write two words that include the root bene. ....

29. Write two words that include the root fac. ....

Using the root information, try to figure out the definitions of the bold word below.

30. We studied the **antebellum** period of the Civil War during history class.

....

31. My mother is a **benefactor** for the local museum.

....

32. There is a **benefit** party for the Boys and Girls Club.

....

Let's get some fitness in! Go to page 237 to try some fitness activities.

Pythagorean Theorem, Part 1

Review:

The Pythagorean Theorem states that for any right triangle - a triangle with a 90-degree angle - the sum of the two sides each squared is equal to the hypotenuse squared. The formula is: $a^2 + b^2 = c^2$ with the variable "c" representing the hypotenuse.

Question 1:

What is the hypotenuse of a right triangle with side lengths of 5 cm and 12 cm?

A. 169 cm

B. 13 cm

C. 17^2 cm

D. $\sqrt{17}$ cm

Question 2:

What is the hypotenuse of a right triangle with side lengths of 9 cm and 12 cm?

A. 15 cm

B. 225 cm

C. 21 cm

D. $\sqrt{15}$ cm

Question 3:

What is the hypotenuse of a right triangle with side lengths of 10 cm and 24 cm?

A. 24 cm

B. 34 cm

C. 34^2 cm

D. 26 cm

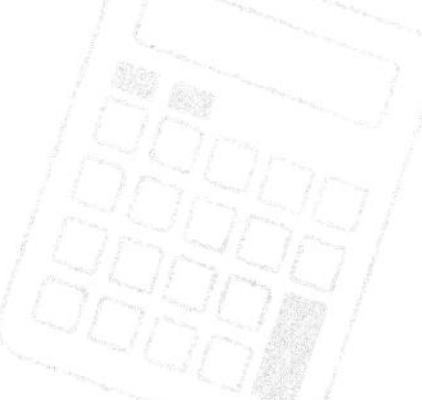

Question 4:

What is the hypotenuse of a right triangle with side lengths of 1 cm and 1 cm?

A. 1 cm

B. 1^2 cm

C. 2 cm

D. $\sqrt{2}$ cm

Question 5:

Mystical Ninja appreciates the mathematic perfection of the Pythagorean Theorem. He draws a right triangle in the sand. The side lengths are 3 meters and 4 meters. What is the length of the hypotenuse?

Question 6:

Lying on the sand on a summer night, Mystical Ninja finds a triangle amidst the constellations. From his perspective, the side lengths measure 8 cm and 15 cm, and the hypotenuse measures 16 cm. Is this a right triangle? Explain.

Question 7:

Mystical Ninja meditates on the ocean waves while lounging on his triangular inflatable raft. It has side lengths of $\sqrt{3}$ and $\sqrt{5}$. Its hypotenuse is the $\sqrt{8}$. Is the raft in the shape of a right triangle? Explain.

Question 8:

Even mathematical geniuses must eat. Mystical Ninja eats a triangular slice of pizza with side lengths of 12 inches and 16 inches and a hypotenuse of 20 inches. Is this pizza slice a right triangle? Explain.

Let's get some fitness in! Go to page 237 to try some fitness activities.

Pythagorean Theorem, Part 2

Review:

Given the lengths of any two sides of a right triangle, the Pythagorean Theorem can be used to find the length of the third side.

Question 1:

What is the length of the missing side in this right triangle?

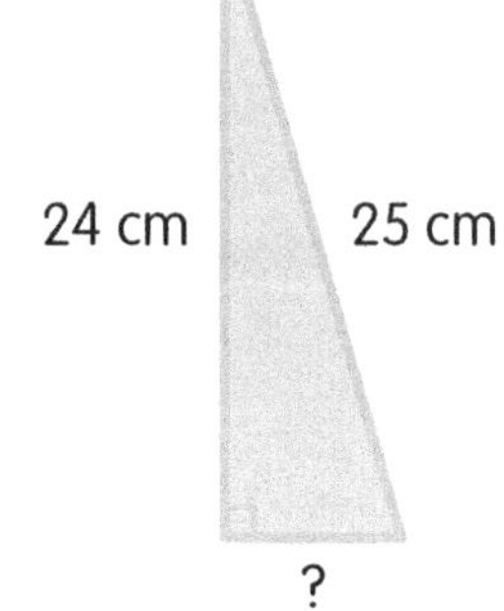

A. 7 cm

B. 49 cm

C. 1,201 cm

D. $\sqrt{1,201}$ cm

Question 2:

What is the length of the missing side in this right triangle?

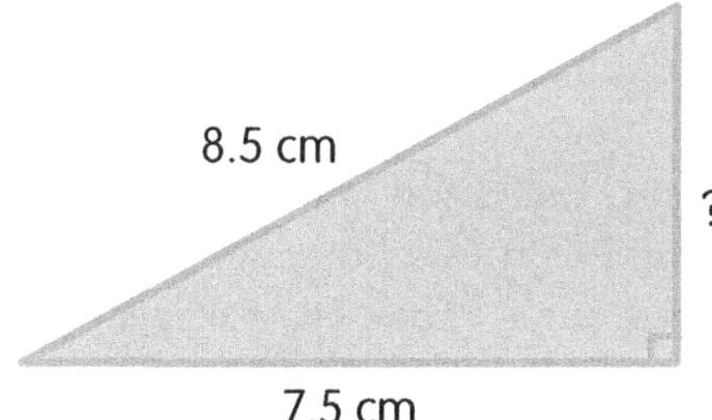

A. 16 cm

B. 4 cm

C. 72.25 cm

D. $\sqrt{72.25}$ cm

Question 3:

What is the length of the missing side in this right triangle?

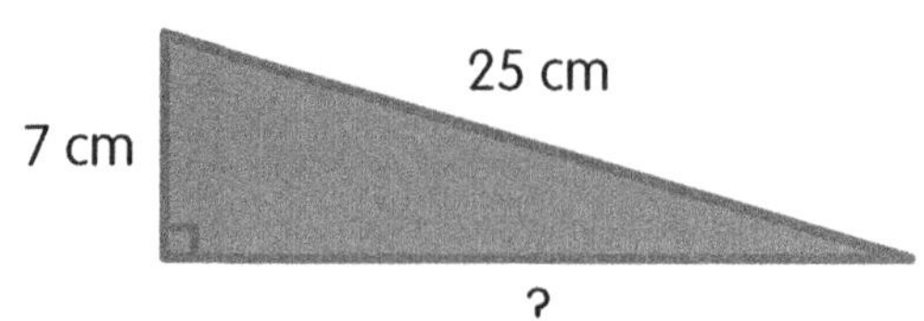

A. 24^2 cm

B. 674 cm

C. 24 cm

D. $\sqrt{674}$ cm

Question 4:

What is the length of the missing side in this right triangle?

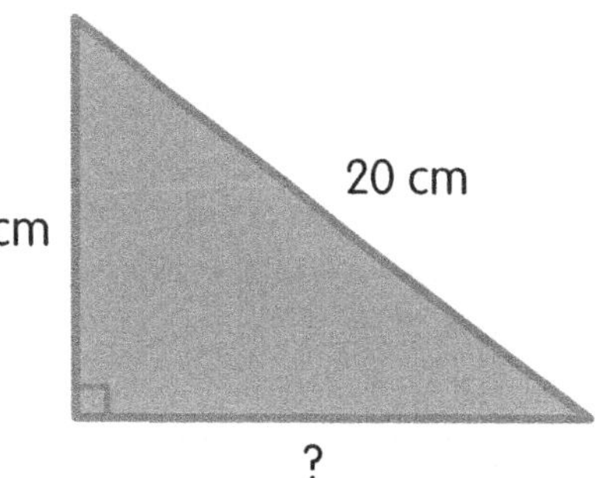

A. 16 cm

B. 16^2 cm

C. 4 cm

D. $\sqrt{32}$ cm

Question 5:

Mystical Ninja has a dream of sailing around the world. He's sewing sails for a sailboat in the shape of right triangles. His main sail has a side length of 5 meters and a hypotenuse of 13 meters. What is the length of the third side?

Question 6:

Mystical Ninja's jib sail has a side length of 4 meters and a hypotenuse of 8 meters. What is the length of the third side?

Question 7:

Mystical Ninja's storm sail has a side length of 2 meters and a hypotenuse of 4 meters. What is the length of the third side?

Question 8:

Mystical Ninja uses an ancient mariner's tool for navigation called a sextant. It is in the shape of a right triangle. The base measures 30 cm and the height measures 15 cm. What is the length of the hypotenuse?

Let's get some fitness in! Go to page 237 to try some fitness activities.

Pythagorean Theorem, Part 3

Review:

The Pythagorean Theorem can be used to find missing lengths of various shapes: the height of a cone, the height of a pyramid, the height of an equilateral triangle, or the length of a square's diagonal.

Question 1:

What is the height of this pyramid?

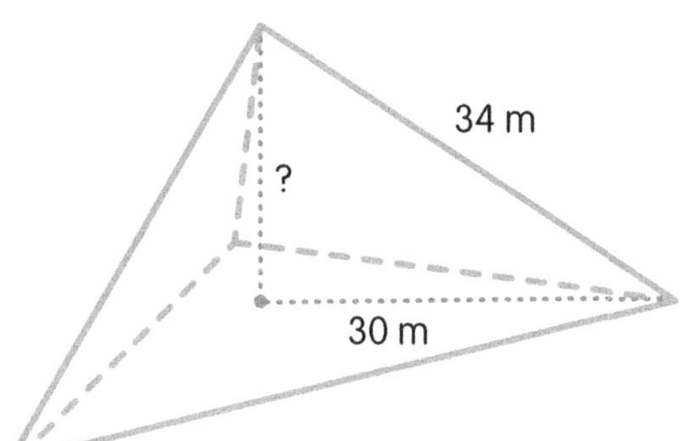

A. 256 meters

B. 16 meters

C. 64 meters

D. 4,096 meters

Question 2:

What is the height of this cone?

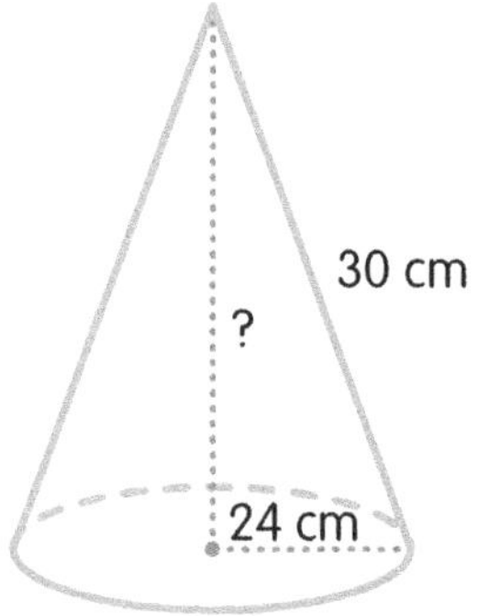

A. 324 cm

B. 18^2 cm

C. 18 cm

D. 54 cm

Question 3:

What is the height of this equilateral triangle?

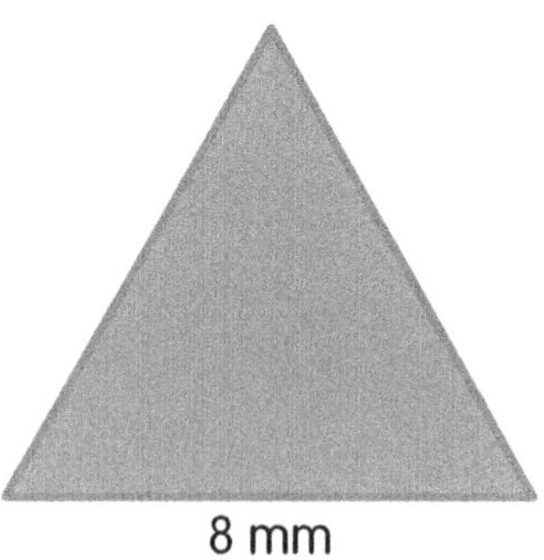

A. $\sqrt{48}$ mm

B. 48 mm

C. 4 mm

D. None of the above

Question 4:

What is the length of the diagonal of this square?

6 cm

A. $\sqrt{36}$ cm

B. 72 cm

C. 6 cm

D. $\sqrt{72}$ cm

Question 5:

Mystical Ninja's dream of sailing the world has come true. He sees a pyramid in Egypt with a sloping side length of 15 hectometers. The distance across the ground from a corner to the center is 12 hectometers. What is the height of the pyramid?

Question 7:

In Brazil, Mystical Ninja sees the side of a house that is a perfect equilateral triangle. He estimates the bottom width of the house to be 10 meters. If this estimate is correct, how tall is the house to the tip of the roof?

Question 6:

Mystical Ninja enjoys a scoop of gelato in Italy on a cone. The cone has a sloping side length of 25 centimeters and diameter of 14 centimeters across the top. How tall is the cone from the bottom tip to the top?

Question 8:

Mystical Ninja walks the perimeter of Tiananmen Square in China. He measures the length of one side to be 500 meters. Assuming it's a perfect square, what would the length of its diagonal be?

Let's get some fitness in! Go to page 237 to try some fitness activities.

Water On The Moon

Recently, NASA scientists have discovered something that baffled them. They found water molecules on the moon. Sure, they have known there is water on the moon for numerous years. This is nothing new. But what is new is that these newly-discovered water molecules were not just found on the far ends on the moon but on the sunlit sides as well. This was very surprising to NASA scientists because they assumed the molecules would not be in that location due to evaporation from the heat of the sun.

Before this discovery, NASA scientists found there was ice on the southern end of the moon. This was interesting to them because this southern tip of ice mimics the North and South poles on the Earth. Just like our planet, the poles on the moon are far away from direct sunlight, so it makes sense that ice resides there. On the moon and on the Earth, the sun does not reach the far polar ends, which causes ice to form. But what surprised scientists was the molecules were found on the sunlit portions of the moon. Why were they there? Why didn't they evaporate from the heat of the sun?

The discovery of the hydrogen and oxygen water molecules prompted several studies and tests. NASA scientists found the water molecules not in frozen ice poles or in large puddles, but instead in separate small molecules on the sunlit surface. The scientists conducted studies to find out where the water molecules came from and why they were not evaporating. Initially, the scientists discovered that the water molecules were trapped inside moon dust. The wind brought the moon dust to the surface, and since the water molecules were trapped, the sun did not melt the molecules.

This shattering discovery adds to the wonderment of thoughts about the moon producing and storing water, which generates more questions, especially regarding the presence of life on the moon.

Another reason why it is baffling the moon has water on its sunlit surface is because the moon is very dry, often comparable to the Sahara Desert on Earth.

1. How do you think the NASA scientists conducted some of these tests? What equipment do you think they used?

2. How do you think scientists discovered water on the moon's sunlit surface?

3. Why is the discovery of water molecules on the sunlit surface of the moon important?

4. How does the water on Earth mimic the water found on the moon?

5. Why do you think the water molecules didn't evaporate from the heat of the sun?

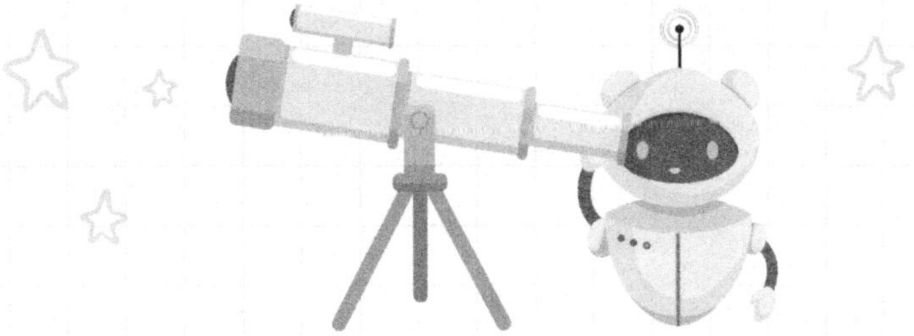

Let's get some fitness in! Go to page 237 to try some fitness activities.

MAZE

Directions:
Connect the tube and toothbrush.

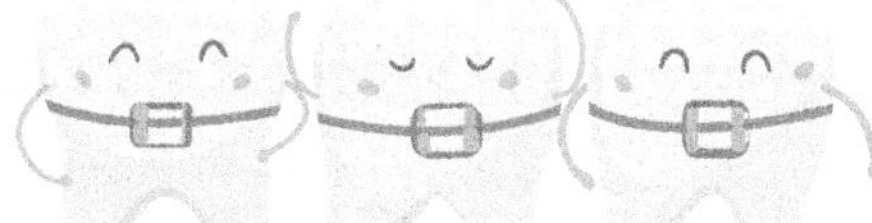

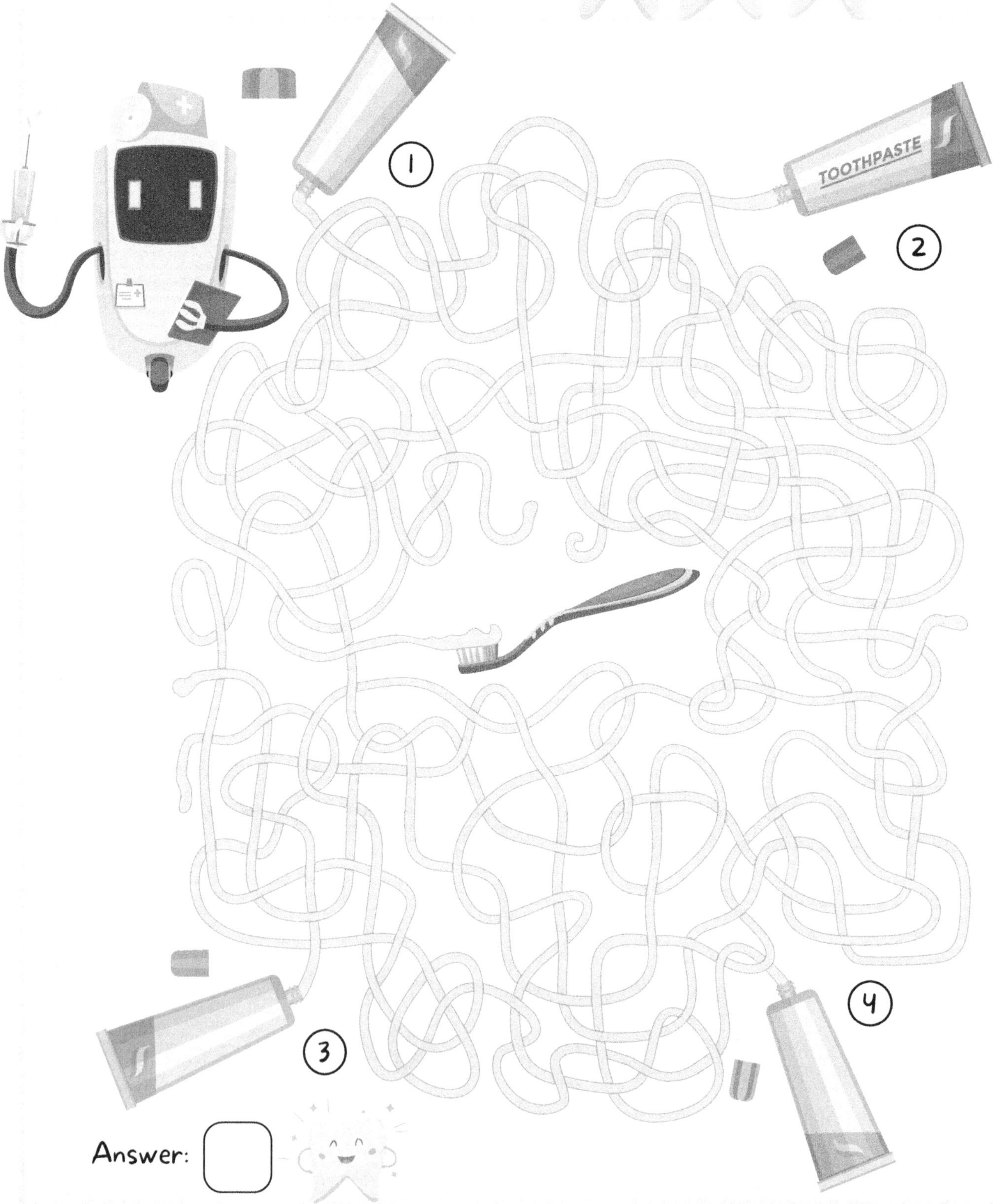

1

TOOTHPASTE

2

3

4

Answer: ☐

WEEK 10

GRADE 8-9

Week 10 dives into synonyms, antonyms, homophones, and sentence fragments. You will also work on math problems dealing with volume of 3-dimensional figures.

Part 1: Synonyms and Antonyms

Synonyms

Synonyms are words that are the same or mean the same thing.
Here is an example - The words happy and cheerful are synonyms because they mean the same thing.
It is important to know synonyms for other words in order to alter your word choice and choose more exciting words that fit the particular mood of the sentence.

Read the following sentences. Then choose a synonym for the underlined word.

1. Terry looks outside at the peculiar view of his surroundings. The sun is in the sky, but there is three feet of snow on the ground. ....

2. My mother told my brother, Steven, that he can have twenty dollars a week for allowance, but she wasn't going to let him borrow money from her anymore. ....

3. In history class today, we watched a movie about the Roman crusades. We had to pay attention to it because there was a pop quiz after it was over. ....

4. My English teacher, Miss Layman, makes us write and reflect after we receive our essay feedback. My last essay got a C-. ....

5. It is prohibited to walk off of the trail at national parks, but Joe Carusco did anyway in order to try to get photographs of wildlife. ....

6. The weather was extremely gloomy. The sun was behind the clouds, it was cold, and it rained the entire day. ....

7. Being a doctor requires someone to be extremely flexible. Doctors see new and different emergencies every day. ....

8. The class was making a commotion during lunch and during quiet time. Their teacher made them miss fifteen minutes of recess. ....

Antonyms

Antonyms are words with opposite meanings.
Here is an example: The words fake and original are antonyms because they mean the opposite of each other.

Read the sentences below. Then in the line provided, rewrite the sentence with an antonym of the underlined word.

9. Our entire family was cramped in the station wagon on the way to Mexico. My legs were so stiff. ....

....

10. The Washington Wombats were victorious at the annual Washington County Championship football game. ....

....

11. The new art project our teacher, Miss Deavers, introduced takes a lot of creativity since we have to come up with new robots. ....

....

12. I can't believe Joey is so calm standing on the free throw line. If Joey makes it, then his team wins. If he doesn't, then his team loses. ....

....

13. I did not know any of the answers on the math worksheet, so I had to go into school early to get help from my math teacher. ....

....

14. Sally was usually ignored and sat alone during lunch, but Bobby decided to reach out to her and be her friend. ....

15. My uncle Louis went to his shop and saw that it was vandalized. There was a lot of damage to the building.

16. My aunt Linda likes to buy old furniture at thrift shops and enjoys restoring and reselling it.

Part 2: Homophones

Homophones are words that sound the same but have different spellings and different definitions. It is important to recognize common homophones so you don't use the wrong word in your writing.

There is a difference between the words do, due, and dew.

1. Write a sentence for each homophone.

There is a difference between the words personal and personnel.

2. Write a sentence for each homophone.

There is a difference between the words your and you're.

3. Write a sentence for each homophone.

There is a difference between the words affect and effect.

4. Write a sentence for each homophone.

....

....

There is a difference between the words allusion and illusion.

5. Write a sentence for each homophone.

....

....

There is a difference between the words their, there, and they're.

6. Write a sentence for each homophone.

....

....

Are these homophones used correctly? Explain.

7. Who is you're best friend? ....

8. The principle is in charge of a school. ....

9. You will find the cereal over their in the last cupboard. ....

10. Their going to Pittsburgh, Pennsylvania for vacation. ....

11. It is there problem, not ours. ....

12. The author makes an illusion to history. ....

Let's get some fitness in! Go to page 237 to try some fitness activities.

Part 1: Sentence Fragments and Run Ons

A **sentence** contains words that are put together in order to deliver a complete thought or a complete message. A complete sentence must contain a subject and a verb.

A **sentence fragment** contains words that do not express a complete thought. A sentence fragment is missing something in order to make it a complete sentence. A sentence fragment is either missing a subject or a verb.

Here is an example of a sentence fragment that is missing the subject - Going to the mall after class.

This sentence is missing the subject. Who is going to the mall after class?

Here is an example of a sentence fragment that is missing the verb - The war statue at the park.

This sentence is missing a verb. What about the war statue at the park? What happened to it? Was it vandalized?

Another error in sentence structure is run-on sentences. **Run-on sentences** are two or more independent clauses joined together. Run-on sentences contain too much information. To fix a run-on sentence, you usually have to separate the independent clauses with a comma or semicolon, or else you should construct separate and complete sentences.

Here is an example of a run-on-sentence - Joey and Maria went to the amusement park Aaron decided to go to the library to study.

Here is how you can fix the above run on sentence with a semicolon - Joey and Maria went to the amusement park; Aaron decided to go to the library to study.

Here is how you can fix the above run on sentence by making two separate and complete sentences - Joey and Maria went to the amusement park. Aaron decided to go to the library to study.

Read the following sentences. Label them as fragments or run on sentences. Then fix them in order to make them complete.

1. While they were gone to the amusement park.

Is this a fragment or a run on? ....

....

2. Jill's blue bike is in front of the school I don't know where Mike's bicycle is.

Is this a fragment or a run on? ....

....

3. Before Angel has to go to bed.

Is this a fragment or a run on? ....

....

4. There are three ways to complete this math problem I don't know how to do it at all.

Is this a fragment or a run on? ....

....

5. Because she went to Jill's house after school without asking her mom.

Is this a fragment or a run on? ....

....

6. The teacher did not make enough copies of the story I had to share with Sara and Lydia.

Is this a fragment or a run on? ....

....

7. Since she woke up really early in the morning.

Is this a fragment or a run on? ....

....

8. My grandfather and my uncle wanted to go fishing they woke up early in the morning they bought worms at the corner market.

Is this a fragment or a run on? ....

....

9. Since she spilled coffee all over her this morning.

Is this a fragment or a run on? ....

....

10. Julia sang loudly everyone could hear her she sounded amazing.

Is this a fragment or a run on? ....

....

11. When Jill arrives at school in the morning.

Is this a fragment or a run on? ....

....

12. The cat fell asleep he didn't see the mouse run across the floor.

Is this a fragment or a run on? ....

....

Part 2: Phrases and Clauses

Appositive phrases are phrases that follow or rename a noun in order to add more description to it. Basically, an appositive phrase adds information to a sentence and is set off by commas.

Here is an example - My friend, Jill, plays with me at recess.

The word Jill is not needed in the sentence, however, it adds information to it. Since it is not needed, it needs to be offset by commas.

In the following sentences, identify the appositive phrases by offsetting them with commas.

1. My uncle an engineer has to move to Utah for his new job.
2. The cheetah the world's fastest animal is found in Africa.
3. Miss Mina my art teacher is gone today.

4. Do not throw out Maji my favorite stuffed dog before you think about giving it to the neighborhood center.

5. The fastest bird the peregrine falcon can travel several hundred miles in a day.

6. My neighbors' dogs Lola and Marin ran away.

Relative Clauses

A **relative clause** is a dependent clause that contains an adjective or a word that describes a noun. Relative clauses give information about a noun or pronoun and usually start with the following words: who, whom, whose, which, or that.

Here is an example - The man, who lives at the end of the road, works at my school.

Fix the following sentences by adding in a relative clause.

7. My mother's name is Sara. She is from Ohio. She likes playing the piano.

....

8. The principal told the students all the rules for the school year. They listened intently.

....

9. The instrumental concert was popular. Many people attended.

....

10. The clouds were blanketing the sky all day. The clouds finally parted for the sun to appear at the time of the game.

....

Let's get some fitness in! Go to page 237 to try some fitness activities.

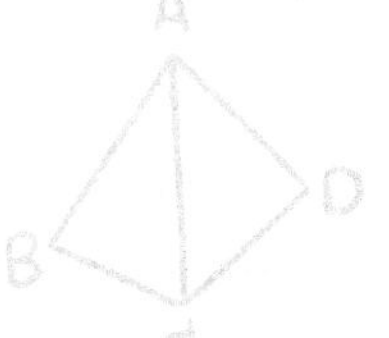

Volume of Cylinders

Review:

The volume of a cylinder can be determined using the formula: $V = \pi \cdot r^2 \cdot h$

In this formula, r is the radius of the circular base and h is the height of the cylinder. Remember that volume is always labeled with units cubed. You may use 3.14 for π in the following problems.

Question 1:

What is the volume of this cylinder?

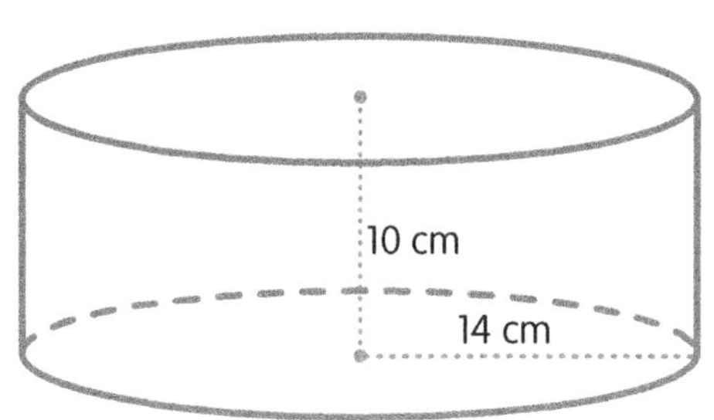

A. 6,154.4 cm^3

B. 140 cm^3

C. 6,154.4 cm^2

D. 439.6 cm^3

Question 2:

What is the volume of this cylinder?

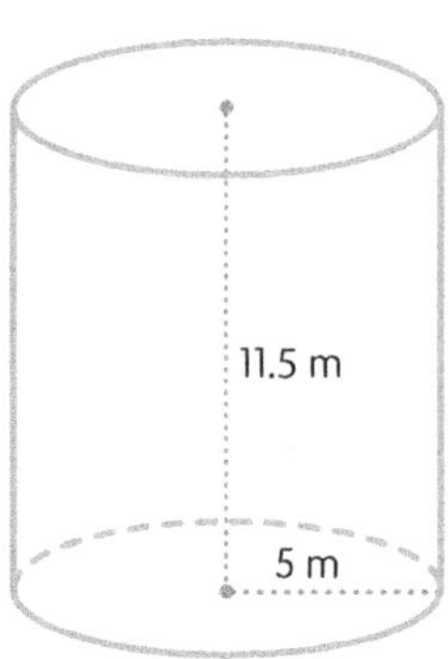

A. 180.55 m^3

B. 287.5 m^3

C. 902.75 m^2

D. 902.75 m^3

Question 3:

What is the volume of this cylinder?

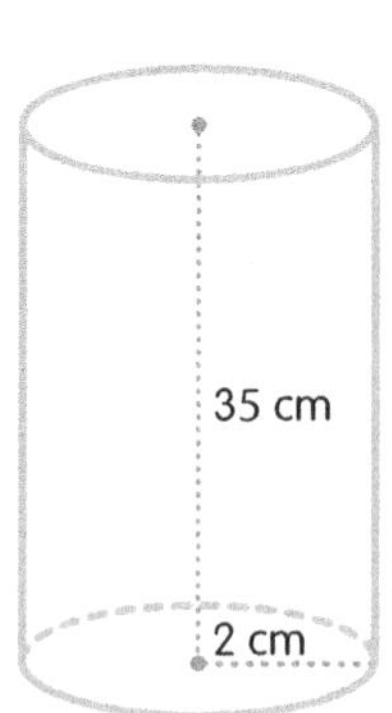

A. 70 cm^3

B. 140 cm^3

C. 439.6 cm^3

D. 109.9 cm^3

Question 4:

What is the volume of this cylinder?

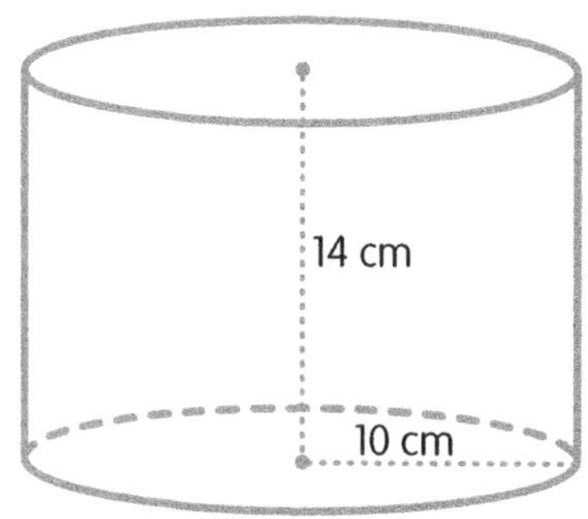

A. 1,400 cm^3

C. 4,396 cm^3

B. 439.6 cm^3

D. 439.6 cm^3

Question 5:

Green Poison uses glass cylinders to display his many ingredients. He has a cylinder for dried jellyfish tentacles with a radius of 10 cm and a height of 49 centimeters. What is the volume of this cylinder?

Question 7:

Green Poison was lucky enough to find a skin shed by a coral reef snake. He displays it in a cylinder with a radius of 25 cm and a height of 50 centimeters. What is the volume of this cylinder?

Question 6:

To hold his collection of barracuda teeth, Green Poison has a glass cylinder with a radius of 8 cm and a height of 20 centimeters. What is the volume of this cylinder?

Question 8:

Oil of pufferfish is one of Green Poison's prized ingredients. He stores it in a glass cylinder with a diameter of 4 cm and a height of 30 centimeters. What is the volume of this cylinder?

Let's get some fitness in! Go to page 237 to try some fitness activities.

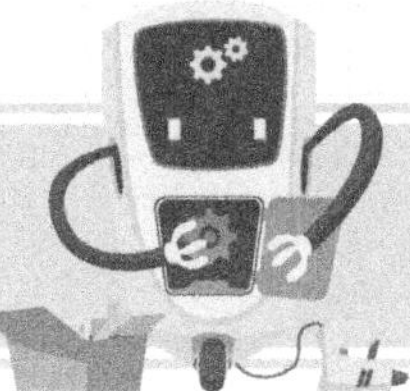

Volume of Cones

Review:

The volume of a cone can be determined using the formula: $V = \frac{1}{3} \cdot \pi \cdot r^2 \cdot h$

In this formula, r is the radius of the circular base and h is the height of the cone. Remember that volume is always labeled with units cubed. You may use 3.14 for π *in the following problems.*

Question 1:

What is the volume of this cone?

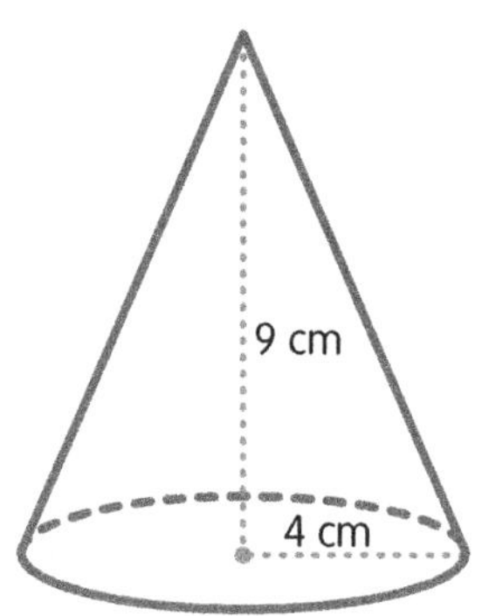

A. 150.72 cm^3

B. 150.72 cm^2

C. 150.72 cm

D. 36 cm^3

Question 2:

What is the volume of this cone?

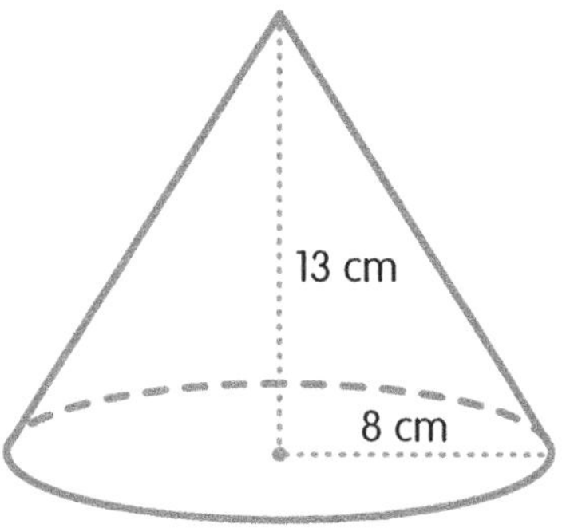

A. 102.4 cm^3

B. 870.83 cm^3

C. 4.92 cm^3

D. None of the above

Question 3:

What is the volume of this cone?

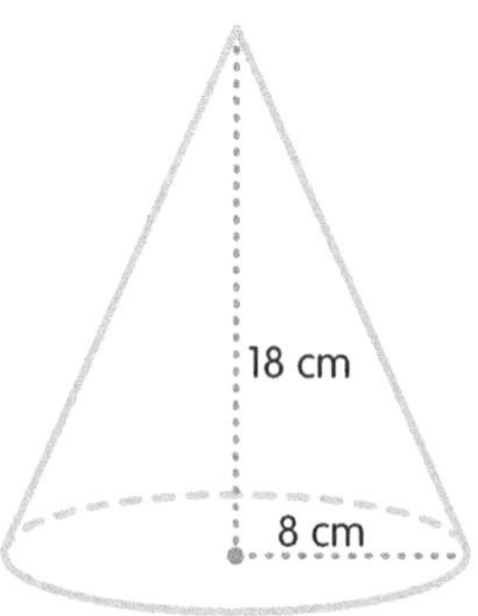

A. 144 cm^3

B. 48 cm^3

C. 56.52 cm^3

D. 1,205.76 cm^3

Question 4:

What is the volume of this cone?

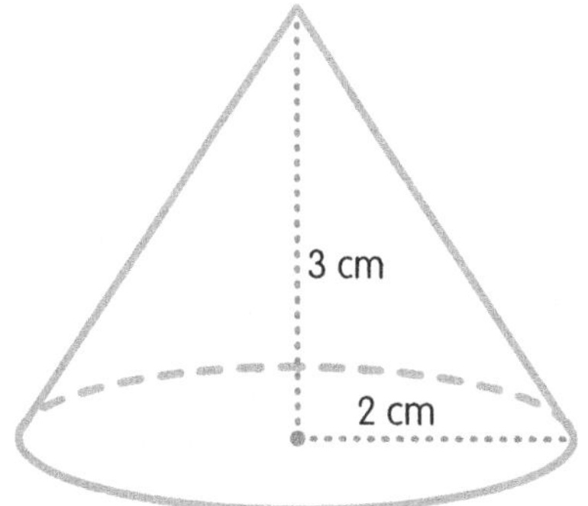

A. 12.56 cm^3

B. 2 cm^3

C. 18.84 cm^3

D. None of the above

Question 5:

Green Poison uses flasks in all shapes and sizes in his chemical laboratory. His favorite is a set of cone-shaped flasks. The tiniest flask has a radius of 2 cm and a height of 4 cm. What is the volume of this flask?

Question 7:

Green Poison dropped his favorite cone-shaped flask, and it busted to pieces. It had a radius of 5 cm and a height of 10 cm. What was the volume of this flask?

Question 6:

The cone-shaped flask Green Poison uses most often has a radius of 6 cm and a height of 15 cm. What is the volume of this flask?

Question 8:

Green Poison was forced to replace his favorite flask with a much larger cone-shaped one. It has a diameter of 10 cm and a height of 20 cm. What is the volume of this flask?

Let's get some fitness in! Go to page 237 to try some fitness activities.

Volume of Spheres

Review:

The volume of a sphere can be determined using the formula: $V = \frac{4}{3} \cdot \pi \cdot r^3$
In this formula, r is the radius of the sphere (the distance from the outside edge to the inner center). Remember that volume is always labeled with units cubed. You may use 3.14 for π in the following problems.

Question 1:

What is the volume of this sphere?

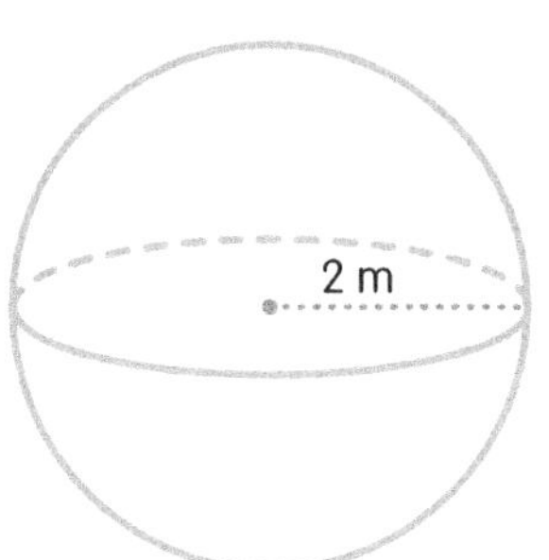

A. 25.12 m^3
B. 50.24 m^3
C. 8.37 m^3
D. 33.49 cm^3

Question 2:

What is the volume of this sphere?

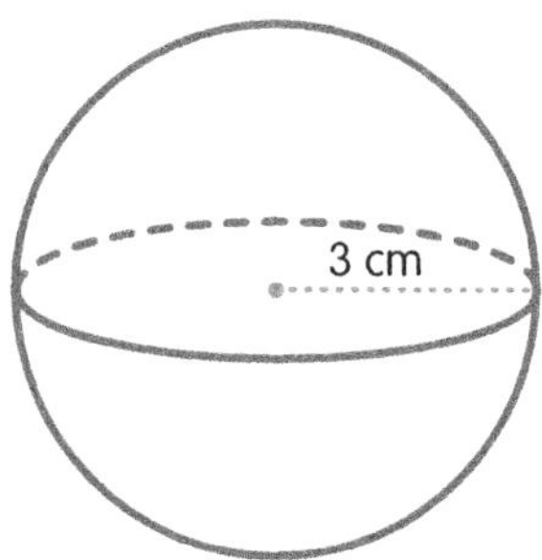

A. 37.68 cm^3
B. 339.12 cm^3
C. 113.04 cm^3
D. 12.56 cm^3

Question 3:

What is the volume of this sphere?

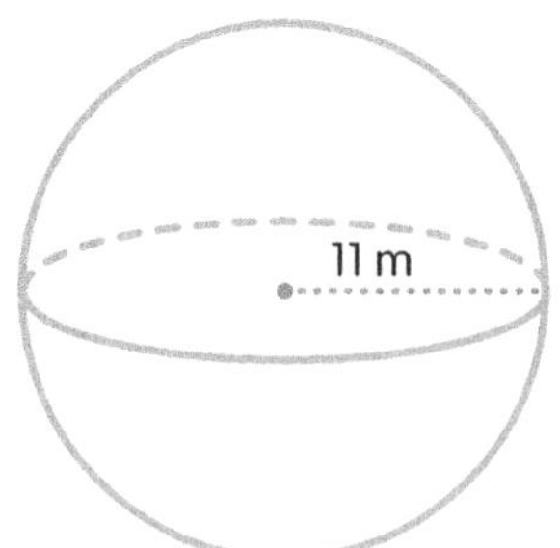

A. 5,572.45 m^3
B. 506.58 m^3
C. 16,717.36 m^3
D. 1,774.67 m^3

Question 4:

What is the volume of this sphere?

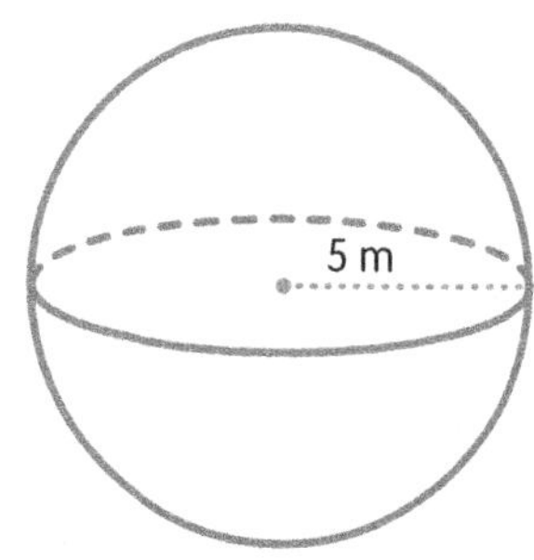

A. $392.5\ m^3$

B. $1,570\ m^3$

C. $104.67\ m^3$

D. $523.3\ m^3$

Question 5:

Green Poison has a fascination with filling glass spheres with colorful toxic substances. He has one small sphere with a radius of 4 cm filled with bright red poison from an urchin. What would be the volume of this sphere?

Question 6:

In a sphere with a radius of 10 cm, Green Poison has the purplish-black ink of a giant squid. What would be the volume of this sphere?

Question 7:

Green Poison treasures his glass sphere of bright blue venom from a stingray. If its radius is 12 cm, what is the volume of this sphere?

Question 8:

In his dreams at night, Green Poison fills a sphere as big as the Earth with fluorescent green poison. Knowing the Earth has a radius of 3958.8 miles, what would be the volume of this sphere?

Let's get some fitness in! Go to page 237 to try some fitness activities.

Have you ever spent time looking at the stars?

A long time ago, before scientists discovered stars and invented the powerful telescopes to see them, people thought that stars were just large diamonds in the sky. Stars have been used to explain religion, legends, and myths. Civilizations have also used them to nagivate to far-off places and to make calendars for planting crops.

Let's conduct an experiment that tests out how much light objects give off and see how this relates to stars.

Materials Needed

Light bulb	Lamp with a removable shade	Measuring tape
Camera stand	Graph paper	Phone or camera

Procedure

1. Turn on a lamp. Make sure the shade has been removed.
2. Set up the camera stand with the camera or phone three feet from the lamp.
3. Make sure all other light sources are off and the window shades are down.
4. Turn on the camera or phone's light. Observe what happens.
5. Measure two feet and move back. Turn on the camera or phone's light.
   Observe what happens.
6. Do the same another two feet away.

Questions and Reflection

1. In your own words, explain what happened as you moved farther away from the lamp?

....

....

2. What is the relationship between light intensity and distance to a light source?

....

....

Discovering Stars
SCIENCE EXPERIMENT

3. What happened when you moved farther away from the light source? What does this mean?

....

....

4. Based on the information from the experiment, what does this tell you about the sun?

....

....

5. Based on the information from the experiment, what does this tell you about the light from the stars?

....

....

6. Do stars produce light? Explain.

....

....

7. Where are the stars in relation to the sun, and how do you know this information?

....

....

Let's get some fitness in! Go to page 237 to try some fitness activities.

FITNESS TIME

MAZE

Directions:

Anchor chains got tangled after the storm. Help the watercrafts find their anchors.

1 ______ 2 ______ 3 ______ 4 ______

WEEK 11

GRADE 8

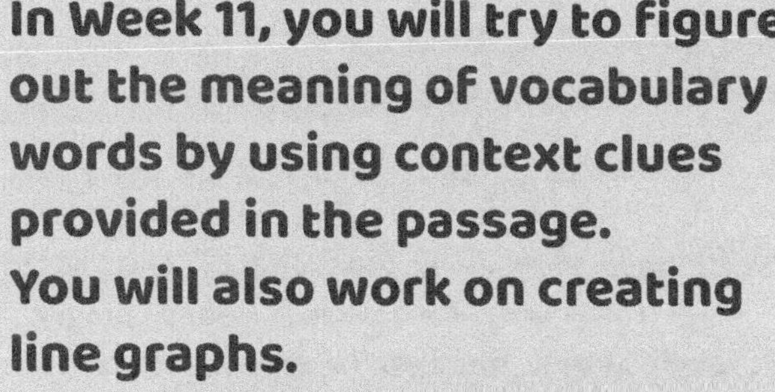

If you are unsure about the meaning of a word and cannot look up the word, then context clues will help you. **Context clues** are clues given in the surrounding text which hint at the meaning of a particular word. The nearby words and phrases will help you figure out what a difficult word means.

Read the nonfiction passage below, paying attention to the words in bold. Then, using context clues or the plug-in strategy, identify what the word means. You can also use the internet to help you if you are stuck!

Eleanor Roosevelt

by Maurine Beasley

(From the permission of the Gilder Lehrman Institute of American History)

Eleanor Roosevelt was one of the most **venerated** women in American history. She was the first lady for twelve years, longer than any other presidential spouse. Through her **endeavors** and her independent spirit, she **revolutionized** the role of first lady from being a hostess to a spokesperson. Eleanor Roosevelt became a role model for millions of Americans due to her **activism** on behalf of social causes like race and gender equality.

Eleanor Roosevelt had a rough childhood, but she grew up to be a very **sovereign** young lady. Before her husband, Franklin D. Roosevelt, became president, Eleanor was a writer and a teacher. She earned her own money and made decisions for herself. Eleanor married her husband, Franklin, when she was twenty years old. At the time, Franklin was finishing his senior year at Harvard University. After graduation, Franklin pursued a political career as assistant secretary of the Navy in Washington. They had six children together. Even though she was **diligent** with her family, she also **immersed** herself in social and domestic **entanglements**.

During World War I, Eleanor volunteered as a Red Cross worker. Doing so made her **self-sufficient,** and she realized that she could build a life for herself outside of her family, just like men. Eleanor recognized **contingencies** open to women after being allowed to vote in 1920. Eleanor formed **alliances** with political women and social reformers. She became well known as a speaker, writer, and a radio **commentator**. She also started a private school for girls in New York City.

When Franklin was elected governor of New York in 1928, there was chatter of him becoming president. At first, Eleanor was **apprehensive** of becoming a first lady because she speculated that she would have to take a backseat in life. She was fearful about going into the White House where she thought her voice would be **stifled**. However, Eleanor did not allow that to occur. Eleanor Roosevelt **metamorphosed** into a figure of importance, not just a president's wife. She soon realized that being first lady enhanced, rather than **constricted,** her career opportunities, and offered her a platform to speak up for the **downtrodden**, women, and racial issues.

Even though some critics felt that Eleanor was overstepping as a first lady, a 1939 Gallup poll found that sixty seven percent of Americans approved of the first lady, even more so than the president. That same year, *Time Magazine* featured Eleanor Roosevelt on its cover and called her the "world's foremost female political force."

After winning his fourth term as president, Franklin died unexpectedly on April 12, 1945. However, Eleanor's career did not come to a halt. She served as a representative to the United Nations and led its Human Rights Commission in developing the Universal Declaration of Human Rights. Today, this document is considered one of the world's most important. Eleanor also continued to write, lecture, and appear on radio and television broadcasts on behalf of liberal causes until she became **fatally** ill from anemia and tuberculosis. She died in New York City on November 7, 1962, at the age of 78. Her death drew front-page coverage all over the world, noting that she "was truly the first lady of the world."

1. What is the definition of the word venerated? ....

2. How do you know? ....

....

3. What is the definition of the word endeavors? ....

4. How do you know? ....

....

5. What is the definition of the word revolutionized? ....

6. How do you know? ....

....

7. What is the definition of the word activism? ....

8. How do you know? ....

....

9. What is the definition of the word sovereign? ....

10. How do you know? ....

....

11. What is the definition of the word diligent? ....

12. How do you know? ....

....

13. What is the definition of the word immersing? ....

14. How do you know? ....

....

15. What is the definition of the word entanglements? ....

....

16. How do you know? ....

....

17. What is the definition of the word self-sufficiency? ....

18. How do you know? ....

....

19. What is the definition of the word contingencies? ....

20. How do you know? ....

....

21. What is the definition of the word alliances? ....

....

22. How do you know? ....

....

23. What is the definition of the word commentator? ....

24. How do you know? ....

....

25. What is the definition of the word apprehensive? ....

26. How do you know? ....

....

27. What is the definition of the word stifled? ....

28. How do you know? ....

....

29. What is the definition of the word metamorphosed? ....

30. How do you know? ....

....

31. What is the definition of the word constricted? ....

32. How do you know? ....

....

33. What is the definition of the word downtrodden? ....

34. How do you know? ....

....

35. What is the definition of the word fatally? ....

36. How do you know? ....

....

Let's get some fitness in! Go to page 237 to try some fitness activities.

Commas

There are many different uses for commas. Here are the most common:

- Commas are used to separate two complete sentences or a complete sentence from a phrase with a coordinating conjunction. Remember that you can use the acronym FANBOYS to recall the coordinating conjunctions (for, and, not, but, or, yet, so).

Here is an example:
Tonya was shy around new adults, but she was never shy around people her own age.

In the example above, the first independent clause is, "Tonya was shy around new adults," and the other independent clause is "she was never shy around people her own age." The two independent clauses are joined by a comma and the coordinating conjunction "but."

- Commas are used after a dependent clause when it comes before a complete sentence. These clauses usually start with the keywords "after," "although," and "because."

Here is an example:
Although Tonya was shy around adults, she was not shy around her peers.

In the above example, the independent clause "she was not shy around her peers" is joined by a comma to a dependent clause that begins with the word "although."

- Commas are used in a series or a list.

Here is an example:
I need to buy bread, milk, soda, butter, and celery at the store.

In the above example, each item is separated by a comma.

Read the sentences below. If a sentence needs a comma, please add it and explain why a comma is needed. If it doesn't need a comma, write "correct" on the line.

1. Even though there was a big mud puddle in the yard and I was wearing new shoes I couldn't resist jumping in it.

....

2. After our team won the state soccer championship I wanted to celebrate.

....

3. In English class we read *Romeo and Juliet To Kill a Mockingbird* and 1984.

4. I completed my math homework jogged around the block fed my cat and played video games.

5. The buck which was the largest deer I ever shot is hanging on the living room wall above the fireplace.

6. While the roast was in a crock pot I baked the apple pie.

7. Under the crawlspace there lived a racoon and her babies.

Semicolons

There are different rules for semicolons. Here are the most common:

- Semicolons connect closely related sentences that do not contain a coordinating conjunction.

Here is an example:
Tonya was shy around adults; she was extroverted around her peers.

Instead of adding a comma and a coordinating conjunction, the independent clause, "Tonya was shy around adults," and the dependent clause, "extroverted around her peers," are joined together by a semicolon.

- Semicolons are used to join two sentences together with a conjunctive adverb. Conjunctive adverbs include the words "consequently," "however," and "therefore."

Here is an example:
Tonya is shy with adults; however, she is not shy with her peers.

The independent clause "Tonya is shy with adults" is connected by a semicolon to a clause that begins with the conjunctive adverb "however."

Read the sentences below. If a sentence needs a semicolon, please add it and explain why a semicolon is needed. If it doesn't need a semicolon, write "correct" on the line.

1. During my winter break, I went to California. My favorite activities were swimming, tanning, and sightseeing.

....

2. Julie is a fantastic hurdler however she fell over three hurdles in the championship meet.

....

3. I want a cute little hamster, but my parents said no. They think hamsters are stinky animals.

....

4. The convict escaped. He went out through the vent in the kitchen.

....

5. I thoroughly enjoyed the novel *Wonder*. It was a pleasure to read.

....

6. The first *Jumanji* movie was incredible. The second one was subpar.

....

7. I hit the snooze button for the last time. It was finally time to get up and get ready for the day.

....

8. I put some ice cubes in my soup. It was way too hot to eat.

....

Colons

There are many different uses for colons. Here are the most common:

- Colons are used to introduce a list or the start of a list.

Here is an example:
Julia has visited the following countries: France, England, Canada, Nigeria, and India.

In the above example, the colon is used before listing off the countries.

- Colons are used like semicolons in order to join sentences together when the second sentence connects or summarizes the first.

Here is an example:
Life is like a board game: half the fun is learning to play.

In the above example, the second sentence summarizes or explains the first and is joined together by a colon.

Read the sentences below. If a sentence needs a colon, please add it and explain why a colon is needed. If it doesn't need a colon, write "correct" on the line.

1. The following students were sent to the principal's office Chloe, Cage, Abby, and Renea.

....

2. There is one thing I expect to see after I went to work and left my dog at home complete chaos.

....

3. These musicals were nominated for a Tony Award *Hamilton, Mean Girls, Dear Evan Hanson,* and *The Color Purple.*

....

4. My mother found a new job at the tutoring center because of her two main qualities her passion for writing and her commitment to students.

....

Let's get some fitness in! Go to page 237 to try some fitness activities.

Interpreting and Creating Line Graphs

Introduction:

A graph tells a story about its data. Measurements across the x-axis and y-axis should be of an appropriate scale and allow for reasonable accuracy.

Question 1:

If this graph depicts the progress of a crab crawling across the beach, what is the story it tells?

A. The crab crawled across the beach at a steady pace for 10 minutes.

B. The crab crawled across the beach, paused, crawled again, paused.

C. The crab crawled slowly across the beach, paused, and then crawled faster.

D. The crab crawled across the beach at a steady pace, paused for several minutes, and then crawled again at the same steady pace.

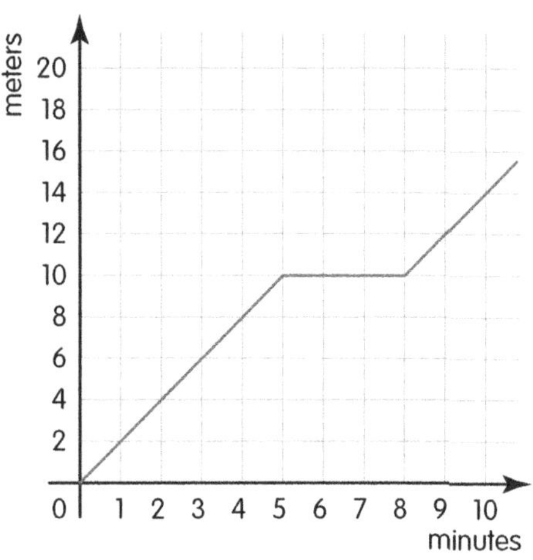

Question 2:

If this graph depicts the progress of an ice cream cone melting in the summer sun, what is the story it tells?

A. The ice cream cone melted at a steady rate over 10 minutes.

B. The ice cream cone melted quickly the first five minutes and slowly the next five minutes.

C. The ice cream cone melted slowly the first three minutes and then faster and faster to be completely melted by the ten-minute mark.

D. The ice cream cone melted quickly the first three minutes and then stopped.

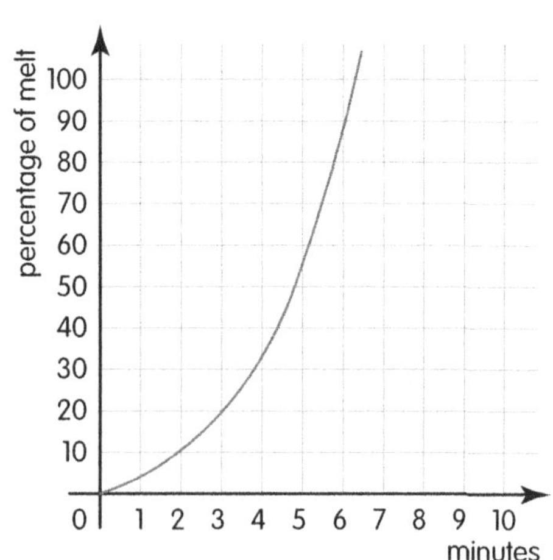

Question 3:

If this graph depicts the elevation change of a scuba diver underwater, what is the story it tells?

A. The scuba diver sunk gradually to -10 meters during the first ten minutes, rose gradually up to -5 meters during the next ten minutes, paused at -5 meters for five minutes, and rose gradually to sea level in the last five minutes.

B. The scuba diver rose from -10 meters to sea level in the first ten minutes, sunk gradually to -5 meters during the next ten minutes, paused at -5 meters for five minutes, and then sunk down to -10 meters for the last five minutes.

C. The scuba diver rose and fell every ten minutes an average of -5 meters.

D. None of the above.

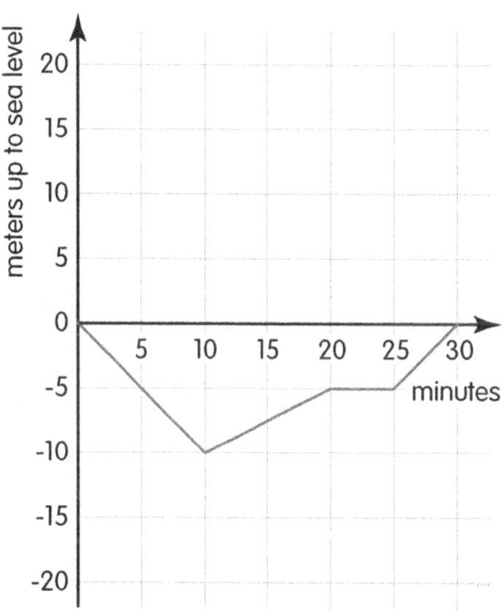

Question 4:

If this graph depicts the change in wind speed during a day on a sailboat, what is the story it tells?

A. The wind was steady and mild for the first half of the day and erratic and strong the entire second half of the day.

B. The wind was steady and mild for the first two hours, rose gradually to a stronger wind that lasted about an hour, rose quickly to an even stronger wind for about an hour, and then dropped quickly back down to no wind by the end of the day.

C. The wind grew strong quickly during the first two hours and then was slow and steady the rest of the day.

D. None of the above

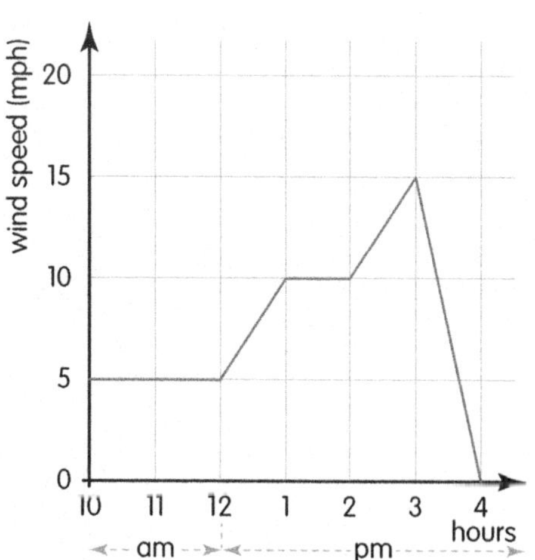

Question 5:

Captain Argo built a sandcastle and spent an hour watching the waves wash it steadily away until there was nothing left. Create a line graph that tells this story with appropriate labels for the x- and y- axis.

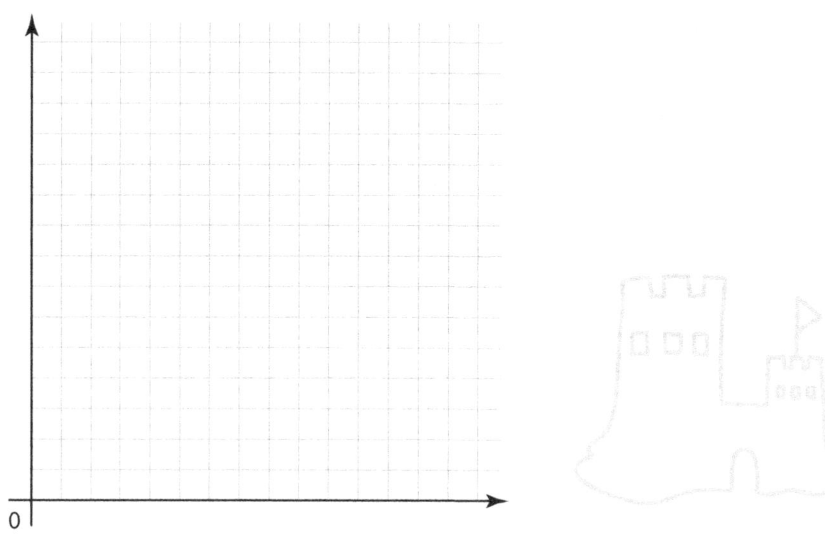

Question 6:

Captain Argo watched the progress of a sea snail across the sand. It moved 10 centimeters over 10 minutes, paused for 10 minutes, and moved 10 more centimeters over 10 more minutes. Create a line graph that tells this story with appropriate labels for the x- and y- axis.

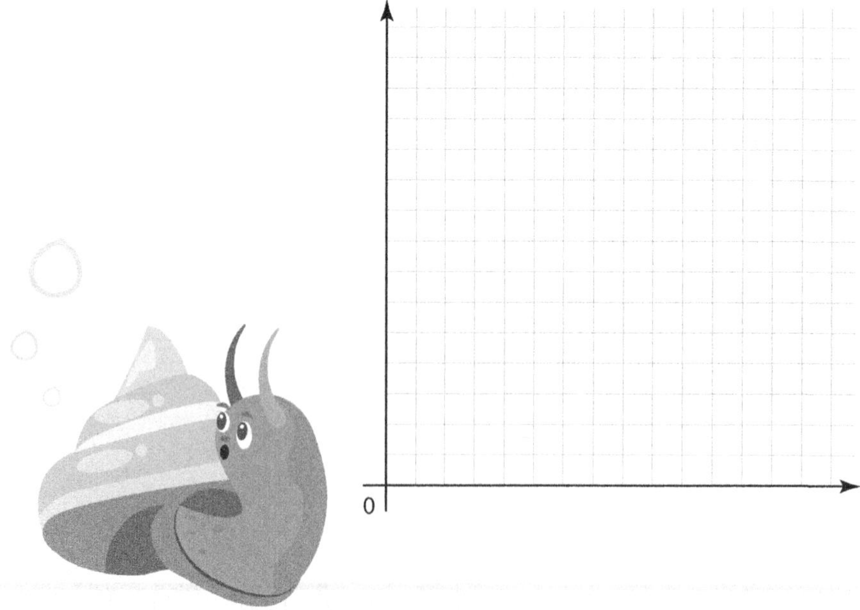

Question 7:

Captain Argo napped in the sun and woke up with a sunburn. He remembers falling asleep for an hour, waking up and not noticing anything at all, and then falling asleep for another hour and waking up with beet-red skin. Create a line graph that tells this story with appropriate labels for the x- and y- axis.

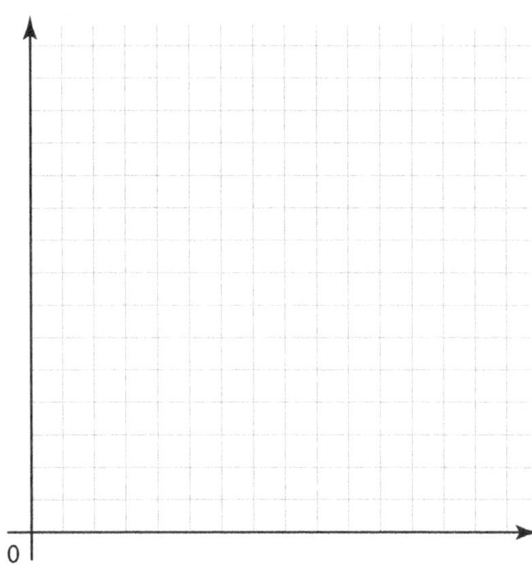

Question 8:

Captain Argo watched the path of the sun during an entire day on the beach. From sunrise to sunset took 12 hours. Sunrise was 6:00 am. The sun was at its highest point at 12:00 pm. Create a line graph that tells this story with appropriate labels for the x- and y- axis.

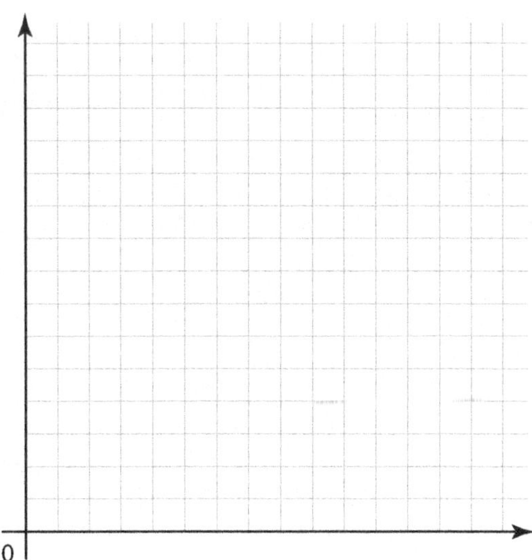

Let's get some fitness in! Go to page 237 to try some fitness activities.

Identifying Parts of an Expression

Introduction:

There are various elements of an algebraic expression. TERMS are the individual parts of the expression separated by addition or subtraction. VARIABLES are symbols used to indicate an unknown number--usually letters of the alphabet. COEFFICIENTS are numbers multiplied by the variable. (If there is no co-efficient, it is assumed to be 1.) A CONSTANT is a term that does not change—meaning it does not have a variable involved.

Question 1:

In the expression $3x + 2y - 5$, what are the individual terms?

A. 3, x, 2, y, and 5.

B. 3x, 2y, and 5.

C. x and y

D. 3, 2, and 5

Question 2:

In the expression $-5a + 7b + c$, what are the variables?

A. -5 and 7

B. a, b, and c

C. -5a, 7b, and 10c

D. None of the above

Question 3:

In the expression $-7x + 2y + z$, what are the coefficients?

A. -7 and 2

B. -7, 2, and 1

C. -7x, 2y, and 5z

D. None of the above

Question 4:

In the expression $2m + 3n - 12$, what is the constant?

A. 2m

B. 3n

C. 12

D. None of the above

Question 5:

Captain Argos is always on the lookout for suspicious terms in algebraic expressions. What terms did he spot in the expression $9a + 10b + 11$?

Question 6:

He may not spot too many villains, but Captain Argos knows a variable when he sees one. What are the variables in the expression $2h + 3i + 4j + 5k$?

Question 7:

Captain Argos doesn't trust coefficients. What are the coefficients in the expression $16m + 12n + 3$?

Question 8:

Captain Argos takes a deep breath when he sees a constant in an algebraic expression. What constant did he see in $6k + 102$?

Let's get some fitness in! Go to page 237 to try some fitness activities.

Simplifying Parts of an Expression

Introduction:

Expressions can sometimes look more complicated than they really are. You can always start to simplify an expression by combining like terms. Variables that are the same (with or without a coefficient) can be combined. Constants can be always be added or subtracted together, too.

Question 1:

Simplify the expression $5c + 8c + 9d - 8$

A. $13c + 9d - 8$

B. $13c + 8d + 17d$

C. $5c + 17d - 8$

D. None of the above

Question 2:

Simplify the expression
$2x + 3x + 5x + 6y + 8y - 9$

A. $5x + 5x + 14y - 9$

B. $5x + 14y - 9$

C. $2x + 3x + 5x + 14y - 9$

D. $10x + 14y - 9$

Question 3:

Simplify the expression
$5a - 3a + 7b - 5b + 3$

A. $5a - 3a + 12b + 3$

B. $2a + 2b + 3$

C. $8a + 12b + 3$

D. None of the above

Question 4:

Simplify the expression $12x - 6x - x + 12$

A. $19x + 12$

B. $5x + 12$

C. $6x - x + 12$

D. $6x + 12$

Question 5:

Captain Argos is a simple guy and likes a simple expression. Simplify the following expression for him: $2a + 2a + 2a - 3b - 3b$.

Question 7:

Captain Argos takes slow, deep breaths and simplifies this expression: $10a - 4a + + 6a - 8 + 7$. What is the simplified expression?

Question 6:

To sleep better at night, Captain Argos knows he needs to tackle this expression: $8x - 4x + 7y - 6y + 20$. What is the simplified expression?

Question 8:

Captain Argos sticks to one outfit most days. It helps him simplify expressions like $8x - 4x - 4x + 7y - 3$. What is the simplified expression?

Let's get some fitness in! Go to page 237 to try some fitness activities.

Buoyancy

How does a boat float?

Many people answer the above question with a simple answer: buoyancy. The definition of the word buoyancy is the ability of something to float in water. But what makes a boat float in a lake and a penny sink to the bottom?

Is the answer that boats are light and that's why they don't sink? Well, a paper boat or a toy boat is light, so it makes sense why it floats on top of the water. But what about a cargo ship or a cruise ship? What makes heavy boats float on the top of the water rather than sink to the bottom?

Archimedes, a Greek scientist, discovered the principle of buoyancy in the year 287 BCE. He actually discovered this while in his bathtub. He noticed that the water level grew based on the size of his body taking up space in the bathtub. He realized that the amount of water displaced would be the size of what entered the water.

Archimedes calculated that the force of an object on water, or a liquid, is equal to the weight of the fluid. Instead of gravity working its magic, the force of buoyancy in water pushes up on an object rather than down. The gravity pulling the boat or ship down is not as great as the buoyancy force pulling up on an object, ultimately making the object float on top rather than sink to the bottom.

Heavy steel ships stay afloat on top of the water because the air is dispersed equally. Even though ships are made of steel, which is a heavy substance, the air in the steel and in other parts of the structure is distributed evenly. The air is less dense than water, which keeps the ship floating. When the weight of the ship equals the volume of the water, then the item, in this case the ship, will float rather than sink.

Questions

1. In your own words, describe the definitions of density and buoyancy?

....

....

2. How was buoyancy discovered?

....

....

3. How does the idea of buoyancy help ship captains load their cargo ships? What do they have to be mindful of?

....

....

4. What if water entered a ship through a hole at the bottom? What would happen? Use the words density, volume, and buoyancy in your answer.

....

....

Let's get some fitness in! Go to page 237 to try some fitness activities.

MAZE

Directions:
Help Brad the monkey get to the zoo employee.

WEEK 12

GRADE 8-9

Congrats on making it to the last week of this workbook! You are truly incredible and your hard work shows it! Go ahead and power through this last week to complete the workbook.

Read the passage and answer the questions that follow.

The Little Match Girl

By Hans Christian Andersen

It was so terribly cold. Snow was falling, and it was almost dark. Evening came on, the last evening of the year. In the cold and gloom a poor little girl, bareheaded and barefoot, was walking through the streets. Of course, when she had left her house she'd had slippers on, but what good had they been? They were very big slippers, way too big for her, for they belonged to her mother. The little girl had lost them running across the road, where two carriages had rattled by terribly fast. One slipper she'd not been able to find again, and a boy had run off with the other, saying he could use it very well as a cradle some day when he had children of his own. And so the little girl walked on her naked feet, which were quite red and blue with the cold. In an old apron she carried several packages of matches, and she held a box of them in her hand. No one had bought any from her all day long, and no one had given her a cent.

Shivering with cold and hunger, she crept along, a picture of misery, poor little girl! The snowflakes fell on her long fair hair, which hung in pretty curls over her neck. In all the windows, lights were shining, and there was a wonderful smell of roast goose, for it was New Year's Eve.

In a corner formed by two houses, one of which projected farther out into the street than the other, she sat down and drew up her little feet under her. She was getting colder and colder, but did not dare to go home, for she had sold no matches, nor earned a single cent, and her father would surely beat her. Besides, it was cold at home, for they had nothing over them but a roof through which the wind whistled, even though the biggest cracks had been stuffed with straw and rags.

Her hands were almost dead with cold. Oh, how much one little match might warm her! If she could only take one from the box and rub it against the wall and warm her hands. She drew one out. *R-r-ratch!* How it sputtered and burned! It made a warm, bright flame, like a little candle, as she held her hands over it; but it gave a strange light! It really seemed to the little girl as if she were sitting before a great iron stove with shining brass knobs and a brass cover. How wonderfully the fire burned! How comfortable it was! The youngster stretched out her feet to warm them too; then the little flame went out, the stove vanished, and she had only the remains of the burnt match in her hand.

She struck another match against the wall. It burned brightly, and when the light fell upon the wall it became transparent like a thin veil, and she could see through it into a room. On the table a snow-white cloth was spread, and on it stood a shining dinner service. The roast goose steamed gloriously, stuffed with apples and prunes. And what was still better, the goose jumped down from the dish and waddled along the floor with a knife and fork in its breast, right over to the little girl. Then the match went out, and she could see only the thick, cold wall. She lit another match. Then, she was sitting under the most beautiful Christmas tree. It was much larger and much more beautiful than the one she had seen last Christmas through the glass door at the rich merchant's home. Thousands of candles burned on the green branches, and colored pictures like those in the print shops looked down at her. The little girl reached both her hands toward them. Then, the match went out. But the Christmas lights mounted higher. She saw them now as bright stars in the sky. One of them fell down, forming a long line of fire. "Now someone is dying," thought the little girl, for her old grandmother, the only person who had loved her, and who was now dead, had told her that when a star fell down, a soul went up to God.

She rubbed another match against the wall. It became bright again, and in the glow the old grandmother stood clear and shining, kind and lovely. "Grandmother!" cried the child. "Oh, take me with you! I know you will disappear when the match is burned out. You will vanish like the warm stove, the wonderful roast goose and the beautiful big Christmas tree!"

And she quickly struck the whole bundle of matches, for she wished to keep her grandmother with her. And the matches burned with such a glow that it became brighter than daylight. Grandmother had never been so grand and beautiful. She took the little girl in her arms, and both of them flew in brightness and joy above the Earth, very, very high, and up there was neither cold, nor hunger, nor fear. They were with God. But in the corner, leaning against the wall, sat the little girl with red cheeks and a smiling mouth, frozen to death on the last evening of the old year. The New Year's sun rose upon a little pathetic figure. The child sat there, stiff and cold, holding the matches, of which one bundle was almost burned.

"She wanted to warm herself," the people said. No one imagined what beautiful things she had seen, and how happily she had gone with her old grandmother into the bright New Year.

1. In paragraph one, the author explains that the little girl is "bareheaded and barefooted." What example of figurative language can be seen in the quote and explain why.

....

....

2. Why did the little girl have matches? Explain.

....

....

3. In paragraph four, the author includes "R-r-ratch" to describe the fire. This is an example of what type of figurative language and why?

....

....

....

4. Is the Christmas tree, goose, and grandmother real? Explain why the author had the little girl view these images.

....

....

....

5. What did the falling star remind the little girl of?

....

....

....

6. The audience only meets the little girl, but the author alludes to the grandmother and the father. Describe them.

....

....

....

7. What is the theme of the story?

....

....

....

8. What is the symbolism in the story and what does the symbol stand for?

....

....

....

9. What is the tone of the story?

....

....

....

10. What do you think would have happened if the little girl went home?

....

....

....

11. Did the girl have another choice than how the story ended?

....

....

....

Let's get some fitness in! Go to page 237 to try some fitness activities.

Read the passage and answer the questions that follow.

Legacy of Rosa Parks

by Barack Obama

(Public Domain Material)

On December 1, 1955, our Nation was forever transformed when an African-American seamstress in Montgomery, Alabama, refused to give up her seat on a city bus to a white passenger. Just wanting to get home after a long day at work, Rosa Parks was not planning to make history, but her defiance spurred a movement that advanced our journey towards justice and equality for all.

Though Rosa Parks was not the first to confront the injustice of segregation laws, her courageous act of civil disobedience sparked the Montgomery Bus Boycott - 381 days of peaceful protest when ordinary men, women, and children sent the extraordinary message that second-class citizenship was unacceptable. Rather than ride in the back of buses, families and friends walked. Neighborhoods and churches formed carpools. Their actions stirred the conscience of Americans. Their resilience in the face of fierce violence and intimidation ultimately led to the desergregation of public transportation systems across our country.

Rosa Parks' story did not end with the boycott she inspired. A lifelong champion of civil rights, she continued to give voice to the poor and the marginalized among us until her passing on October 24, 2005.

As we mark the 100th anniversary of Rosa Parks' birth, we celebrate the life of a genuine American hero and remind ourselves that although the principle of equality has always been self-evident, it has never been self-executing. It has taken acts of courage from generations of fearless and hopeful Americans to make our country more just. As heirs to the progress won by those who came before us, let us pledge not only to honor their legacy, but also to take up their cause of perfecting our Union.

I, Barack Obama, President of the United States of America, by virtue of the authority vested in me by the Constitution and the laws of the United States, do hereby proclaim February 4, 2013, as the 100th Anniversary of the birth of Rosa Parks. I call upon all Americans to observe this day with appropriate service, community, and educational programs to honor Rosa Parks' enduring legacy.

1. What did Parks do on December 1, 1955 that she is still remembered for today?

A. On December 1, 1955, Rosa Parks gave up her seat on the bus to a Caucasian individual.

B. On December 1, 1955, Rosa Parks refused to give up her bus seat to a Caucasian individual.

C. On December 1, 1955, Rosa Parks participated in the Montgomery Bus Boycott in the state of Alabama.

D. On December 1, 1955, Rosa Parks celebrated her birthday.

2. What happened due to Rosa Parks' disobedience?

A. Due to Rosa Parks' disobedience, all African Americas riding busses were allowed to sit wherever they wanted.

B. Due to Rosa Parks' disobedience, African Americans opened their eyes to the racism surrounding them.

C. Due to Rosa Parks' disobedience, African Americans participated in the Montgomery Bus Boycott.

D. Due to Rosa Parks' disobedience, her birthday became a national holiday.

3. In Barack Obama's speech, it reads "Ordinary men, women, and children sent the extraordinary message that second-class citizenship was unacceptable." What was that message and what was its intention?

....

....

....

....

4. What was the purpose of President Barack Obama giving this speech about the legacy of Rosa Parks?

A. Barack Obama wanted to remind the citizens of the United States that African Americans faced many injustices.

B. Barack Obama wanted to teach the citizens of the United States about the life of Rosa Parks.

C. Barack Obama wanted to encourage the citizens of the United States about the civil disobedience that occurred.

D. Barack Obama wanted to honor Rosa Parks and have all Americans observe her birthday.

5. Read the following sentence from Barack Obama's speech: "Just wanting to get home after a long day at work, Rosa Parks was not planning to make history, but her defiance spurred a movement that advanced our journey toward justice and equality for all." What does the word defiance mean?

A. The word defiance means to stop something from occurring.

B. The word defiance means to get the permission to do something.

C. The word defiance means to refuse to obey someone or to refuse to do something.

D. The word defiance means the act of obeying someone's authority.

6. What did President Barack Obama want the citizens of the United States of America to do on February 4, 2013?

....

....

....

7. What was the Montgomery Bus Boycott?

....

....

....

8. What was the purpose of the Montgomery Bus Boycott?

....

....

....

Let's get some fitness in! Go to page 237 to try some fitness activities.

The Distributive Property

Introduction:

The distributive property of multiplication is used to write equivalent expressions. This rule in algebra states:

$$a(b + c) = a \cdot b + a \cdot c$$

Question 1:

Using the distributive property, what is the equivalent expression for $3(5a + 3b)$?

A. $15a + 9b$
B. $8a + 6b$
C. $15a + 3b$
D. $5a + 9b$

Question 2:

Using the distributive property, what is the equivalent expression for $2(5a + 5)$?

A. $10a + 5$
B. $10a + 10$
C. $5a + 10$
D. $10a$

Question 3:

Using the distributive property, what is the equivalent expression for $8(a + b)$?

A. $8a + b$
B. $8 + a + b$
C. $8a + 8b$
D. $16a + 16b$

Question 4:

Using the distributive property, what is the equivalent expression for $5(a + 7b + 10)$?

A. $5a + 7b + 50$
B. $5a + 7b + 10$
C. $a + 35b + 10$
D. $5a + 35b + 50$

Question 5:

Adrastos the Super Warrior has just finished a busy day of tv interviews. He settles down at home, admires himself in the mirror, and mulls over the equivalent expression for $10(2a + 6b + 7c + 10)$. Using the distributive property, what is the equivalent expression?

Question 7:

Taking selfies on his phone, Adrastos the Super Warrior poses and decides on the equivalent expression for $4(x - 3 + y)$. What is this expression?

Question 6:

As he autographs photos of himself, Adrastos the Super Warrior works out the equivalent expression for $5(a + b - 12)$. Using the distributive property, what is the equivalent expression?

Question 8:

Adrastos the Super Warrior examines his collection of 8 dozen capes. He settles on one and also chooses the equivalent expression for $2(a - 7 - 9b - 3c)$ Using the distributive property, what is the equivalent expression?

Let's get some fitness in! Go to page 237 to try some fitness activities.

The Commutative Property

Introduction:

The commutative property is used to write equivalent expressions. It states the order of numbers on which we operate can be moved or swapped around. This rule in algebra holds for the operations of addition and multiplication but not subtraction or division. This rule states:

$$a + b = b + a$$
and
$$a \cdot b = b \cdot a$$

Question 1:

Using the commutative property, what is an equivalent expression for $6a + 10b$?

A. $10b + 6a$
B. $6a - 10b$
C. $10b - 6a$
D. None of the above

Question 2:

Using the commutative property, what is an equivalent expression for $12x \cdot 12y$?

A. $12y \cdot 12x$
B. $24(x + y)$
C. $24xy$
D. All of the above

Question 3:

Using the commutative property, what is an equivalent expression for $2a + 2b + 3c$?

A. $3c + 2a + 2b$
B. $3c + 2b + 2a$
C. $2a + 3c + 2b$
D. All of the above

Question 4:

Using the commutative property, what is an equivalent expression for $6x \cdot 5y \cdot 3z$?

A. $3z \cdot 5y \cdot 6x$
B. $5y \cdot 3z \cdot 6x$
C. $6x \cdot 3z \cdot 5y$
D. All of the above

Question 5:

Adrastos the Super Warrior is all about community. He hands out autographs and shares his knowledge about the commutative property of algebra everywhere he goes. What does he say is an equivalent expression for $8a + 7$?

Question 6:

Adrastos the Super Warrior knows everyone appreciates a clear example of the commutative property. What are five other equivalent expressions for $19 + 2x + 5y$?

Question 7:

Adrastos the Super Warrior loves to lecture his fans on the value of recycling and the commutative property. What does he say are the three other equivalent expressions for $6x \cdot 6y \cdot 6z$?

Question 8:

Adrastos the Super Warrior likes to challenge his audience to pick up trash and know the commutative property by heart. He says there are many equivalent expressions for $a + b + a \cdot b$. What are the four possibilities?

Let's get some fitness in! Go to page 237 to try some fitness activities.

The Associative Property

Introduction:

The associative property is used to write equivalent expressions. It states the sum or product of a set of numbers is the same no matter how they are grouped. This rule in algebra holds for the operations of addition and multiplication but not subtraction or division. This rule states:

$$(a + b) + c = a + (b + c)$$
and
$$(a \times b) \times c = a \times (b \times c)$$

Question 1:

According to the associative property, what is an equivalent expression for (5a + 7b) + 8c?

A. 5a x (7b) + 8c
B. 5a + (7b + 8c)
C. (5a x 7b x 8c)
D. 5a x (8c + 7b)

Question 2:

According to the associative property, what is an equivalent expression for 10x + (11y + 12z)?

A. (10x + 11y) + 12z
B. 12z + 11y x 10x
C. (10x · 12z · 11y)
D. 10x(11y + 12z)

Question 3:

According to the associative property, what is an equivalent expression for (5a x 7b) x 8c?

A. 5a x (7b x 8c)
B. 5a + (7b x 8c)
C. (5a x 7b + 8c)
D. (5a + 8c) x 7b

Question 4:

According to the associative property, what is an equivalent expression for 10m x (11n x 12p)?

A. (10m x 11n) + 12p
B. 12p + (11n x 10m)
C. (10m x 11n) x 12p
D. None of the above

Question 5:

Adrastos the Super Warrior knows he adds something special to the world just like the associative property adds to the world of algebra. He uses the expression $(2a + 3b) + 4c$ as an example. Using the associative property, what is an equivalent expression?

Question 6:

Not for a moment has Adrastos the Super Warrior questioned his own extraordinary value. He uses the associative property as his proof. He points to the equivalent expression for $(7m + 12n) + 13q$ as an example. Using the associative property, what is an equivalent expression?

Question 7:

Adrastos the Super Warrior periodically throws himself an awards party. He has certificates of achievement made with the associative property stated on the back. He uses $(2a \times 3b) \times 4c$ as an example. Using the associative property, what is an equivalent expression?

Question 8:

Adrastos the Super Warrior only associates himself with superstars and the associative property, of course. He points to the equivalent expression for $(7m \times 12n) \times 13q$ as an example. Using the associative property, what is an equivalent expression?

Let's get some fitness in! Go to page 237 to try some fitness activities.

Composting

Do you recycle at home? What about composting? Composting is similar to recycling in that instead of throwing away natural ingredients in a landfill and letting them decay and rot over time, you can use the waste to help fertilize plants and flowers in your own lawn or garden. Scientists have realized that natural food ingredients, grasses, and leaves decompose in a compost pile, but what about paper and certain plastic products?

Let's test out an experiment in order to find out.

Materials Needed

Paper bowl	Plastic fork, spoon, or knife	Chip bag
Scissors	Large plastic bin	Soil

You will also need some of these items:

fruits, vegetables, dead plants or flowers, weeds, coffee grounds, tea bags, grass cuttings, leaves, newspaper, egg shells, tissues, coffee filters

Procedure

1. Take the paper bowl, plastic silverware, and chip bag and break them up into smaller pieces.
2. Set the large plastic bin outside.
3. Fill it halfway with soil.
4. Add in compost ingredients from the list provided above.
5. Add the cut paper bowl, plastic silverware, and chip bag.
6. Mix the compost and put the lid on it.
7. After a week, come back to it and mix it up again. Make sure you break up any clumps that have formed. Continue to do this every week for a month.
8. At the end of the month, open the compost bin and see what items have not decomposed.
9. After answering the questions, use the compost soil in a potted plant, in your yard, or in your garden.

Questions and Reflection

1. Why did the directions tell you to cut the paper bowl, the plastic silverware, and the chip bag into smaller pieces?

....

....

2. What if the paper bowl, plastic silverware, and chip bag were not cut down? Would this have changed the results?

....

....

3. As the month went on, what did you notice about the temperature and the humidity of the compost bin?

....

....

4. What happened to the volume of the compost pile after the month duration?

....

....

5. What happened to the three items you tested? Are they compostable? Explain.

....

....

....

....

Let's get some fitness in! Go to page 237 to try some fitness activities.

FITNESS TIME

MAZE

Directions: Help the robot Stan to pass the maze to return the book to the library.

GRADE 8-9

FITNESS TIME

FITNESS TIME

exercises complex one

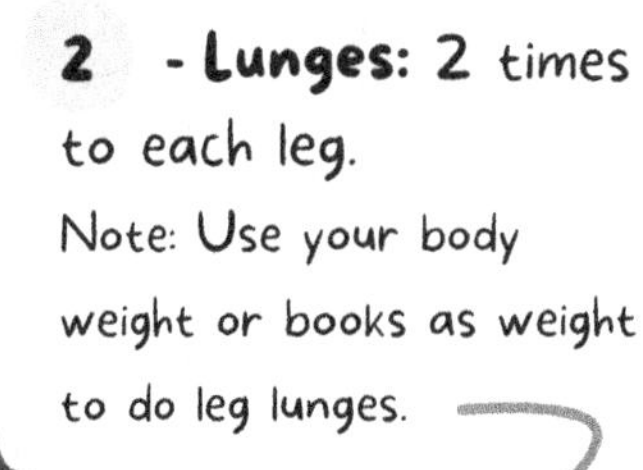

1 - Abs: 3 times

2 - Lunges: 2 times to each leg.
Note: Use your body weight or books as weight to do leg lunges.

3 - Plank: 6 sec.

4 - Run: 50m
Note: Run 25 meters to one side and 25 meters back to the starting position.

"Please be aware of your environment and be safe at all times. If you cannot do an exercise, just try your best."

exercises complex two

1 - High Plank: 6 sec.

2 - Chair: 10 sec.
Note: sit on an imaginary chair while keeping your back straight.

3 - Waist Hooping: 10 times. Note: if you do not have a hoop, pretend you have an imaginary hoop and rotate your hips 10 times.

4 - Abs: 10 times

FITNESS TIME

exercises complex three

2 - **Bend Down:** 10 sec.

3 - **Chair:** 10 sec.

1 - **Down Dog:** 10 sec.

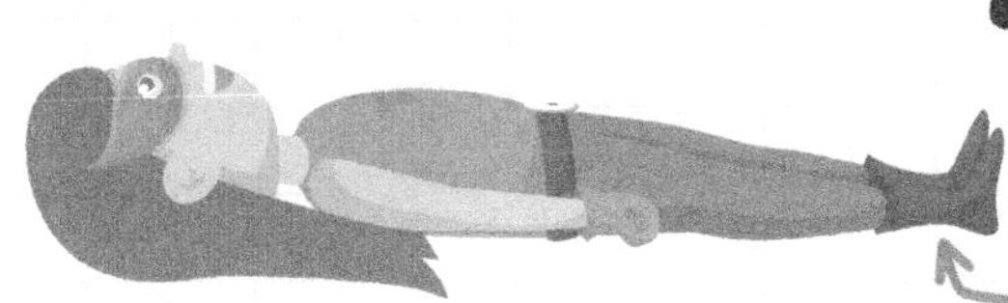

5 - **Shavasana:** as long as you can.
Note: think of happy moments and relax your mind.

4 - **Child Pose:** 20 sec.

Please be aware of your environment and be safe at all times.
If you cannot do an exercise, just try your best.

exercises complex four

2 - **Lunges:** 3 times to each leg.
Note: Use your body weight or books as weight to do leg lunges.

1 - **Bend forward:** 10 times.
Note: try to touch your feet. Make sure to keep your back straight, and if needed, you can bend your knees.

3 - **Plank:** 6 sec.

4 - **Abs:** 10 times

Answer Sheets

To see the answer key to the entire workbook, you can easily download the answer key from our website!

*Due to the high request from parents and teachers, we have removed the answer key from the workbook so you do not need to rip out the answer key while students work on the workbook.

All you need to do is:

Step 1 - Visit our website at: **www.argoprep.com/books**

Step 2 - Choose the workbook you have and you will see **DOWNLOAD ANSWER SHEETS** button as well as all video explanations.

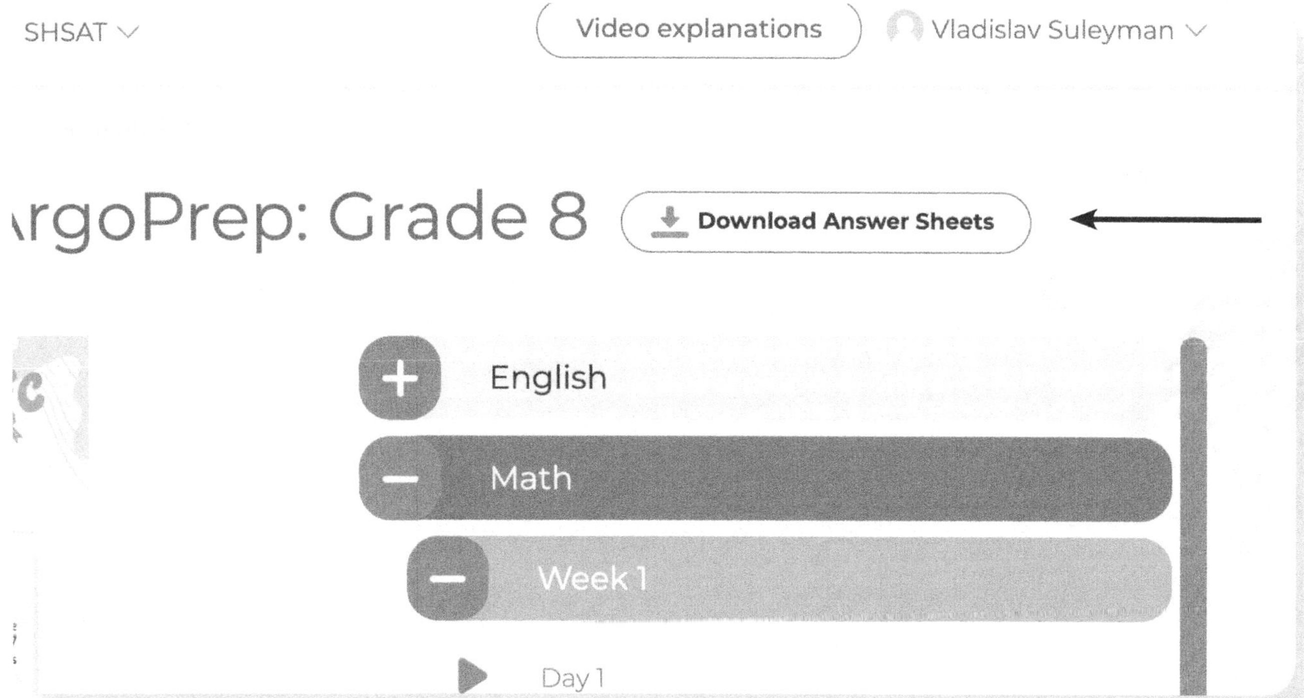

Made in the USA
Monee, IL
19 May 2021

68992243R00136